Laboratory Anatomy of the

Fetal Pig

Ninth Edition
Laboratory Anatomy of the
Fetal Pig

Theron O. Odlaug
University of Minnesota

Skeleton of the Cat

Sheep Heart and Brain
Robert B. Chiasson, University of Arizona

WCB **Wm. C. Brown Publishers**

Book Team

Editor *Kevin Kane*
Developmental Editor *Margaret J. Manders*
Production Coordinator *Peggy Selle*

WCB **Wm. C. Brown Publishers**

President *G. Franklin Lewis*
Vice President, Publisher *George Wm. Bergquist*
Vice President, Operations and Production *Beverly Kolz*
National Sales Manager *Virginia S. Moffat*
Group Sales Manager *Vincent R. Di Blasi*
Vice President, Editor in Chief *Edward G. Jaffe*
Marketing Manager *Paul Ducham*
Advertising Manager *Amy Schmitz*
Managing Editor, Production *Colleen A. Yonda*
Manager of Visuals and Design *Faye M. Schilling*
Production Editorial Manager *Julie A. Kennedy*
Production Editorial Manager *Ann Fuerste*
Publishing Services Manager *Karen J. Slaght*

WCB Group

President and Chief Executive Officer *Mark C. Falb*
Chairman of the Board *Wm. C. Brown*

Cover design Sailer & Cook Creative Services

Copyedited by Julie Bach

The credits section for this book begins on page 108, and is considered an extension of the copyright page.

Contents

List of Figures

Preface

The changes between the 7th and 8th editions of the Fetal Pig manual constituted a major revision of some of the material. In the present 9th edition some of the terminology has been improved by substituting better or corrected terms. A new feature is the statement of chapter objectives followed by an introduction to each chapter. In the chapter on the circulatory system, a flow chart has been added comparing fetal to adult circulation. There has been some rearrangement of certain figures in order to bring them into closer relationship with each other: photomicrographs of the kidney cortex have been moved closer to the drawing of the sagittal section of the kidney. In addition, a new drawing has been added, that of the nephron of the kidney. Figures of the external anatomy of the heart have been moved closer to the figures of the internal anatomy of the heart. Figure 1 (Anatomical planes and directions) has been removed from the chapter on external anatomy and placed with the section on anatomical terminology. Color has been added to Figure 7.2 (Fetal pig heart), Figure 7.5 (Blood vessels of the hepatic portal system), Figure 7.11 (Internal view of the fetal pig heart), and Figures 9.6 and 9.9 (Brain).

Grateful appreciation is expressed to the following for permission to use various figures or for providing drawings or photographs:

Mrs. Lillie C. Grossman: Skeleton of the Pig
Dr. Robert B. Chiasson: Skeleton of the Cat
Carl Petterson: External Features and Fetal Pig Skeleton
Joan Beck: Incisions to Be Made
Mark Summers: Cross Section of the Ileum
Kenneth Moran: Photographs of Dissections of the Fetal Pig
Dr. Arlen R. Severson: Photomicrographs
Dr. John Raynor: Structure of the Ear, Structure of the Inner Ear, Structure of the Cochlea
Dr. John W. Hole, Jr.: Touch and Pressure Receptors, Olfactory Receptors, Taste Buds
Leanne Witzig: Drawings of Organ Systems of the Fetal Pig

The information on fetal pig preparation was provided courtesy of Sargent-Welch Scientific Company, Carolina Biological Supply, MacMillan Scientific Company, and Nasco.

I would also like to express my appreciation to the following reviewers whose comments and suggestions contributed to the development of this edition:

Katherine Jacobson
Southern Illinois University–Carbondale

Donna M. Van Wynsberghe
University of Wisconsin–Milwaukee

Hugh Lefcort
Oregon State University

Also, I wish to acknowledge with thanks the excellent work by the staff of Wm. C. Brown Company in the preparation and production of the Fetal Pig manual, especially that of Marge Manders, Developmental Editor for Biology.

Note: Also available from the publisher is a *Dissection Guide for the Fetal Pig,* which can be purchased as slides (#B10–264–S, $29.95) or as a filmstrip (#B10–265–F, $19.95). To order, call 1–800–338–5578 between the hours of 7:00 A.M. and 6:00 P.M., CST.

Introduction

The pig is a member of the group of mammals known as even-toed ungulates; that is, they possess two toes, the third and fourth. The first toe is lost and the second and fifth are reduced in size or lost. Taxonomically, the pig belongs to the phylum Chordata, class Mammalia, order Artiodactyla, genus *Sus,* species *scrofa.*

The specimens you will dissect and study are fetal (unborn) pigs. Fetal pigs are obtained from pregnant sows in the process of being slaughtered at abatoirs or slaughterhouses. The period of gestation or development within the uterus of the female is 112 to 115 days and there are, on the average, 7 to 12 offspring in a litter. Laboratory specimens are usually 8 to 12 inches in length with the developmental age of the pig between 100 days for the 8-inch length to near full term for the larger ones. The fetuses are removed from the uterus, cooled immediately, and then embalmed with formaldehyde or a phenol-based solution. The embalming is done through one of the umbilical arteries.

Liquid red latex is injected under pressure into the arterial system through one of the umbilical arteries; blue latex is injected in the same manner into the venous system through the external jugular vein. The arteries fill rather completely but this is not always the case with the veins. Veins have much thinner walls than do the arteries and considerable care must be taken when injecting the veins to avoid rupturing them. The cranial veins generally fill fairly well but those caudal to the kidneys are frequently empty of latex.

The fetal pig is an excellent representative of the class of mammals since it illustrates the basic anatomy of the adult mammal as well as providing information about certain features and structures present in the fetal body. Because of the immature nature of the organ systems, however, considerable care must be taken in dissection.

The directions for dissection and the numerous drawings and photographs are for your guidance and help. Read and understand the directions before you begin to dissect, and refer to the manual frequently as you work. Do your own work but do not hesitate to ask for assistance if you run into difficulties. Refer to page xiii for a glossary of anatomical terms which are used throughout the text.

For most effective use, dissecting instruments such as scissors and scalpel should be kept sharp. These instruments, as with all dissecting equipment, must be used with care to prevent injury to the user. Any cut or nick must be promptly sterilized. The wearing of rubber gloves is strongly recommended.

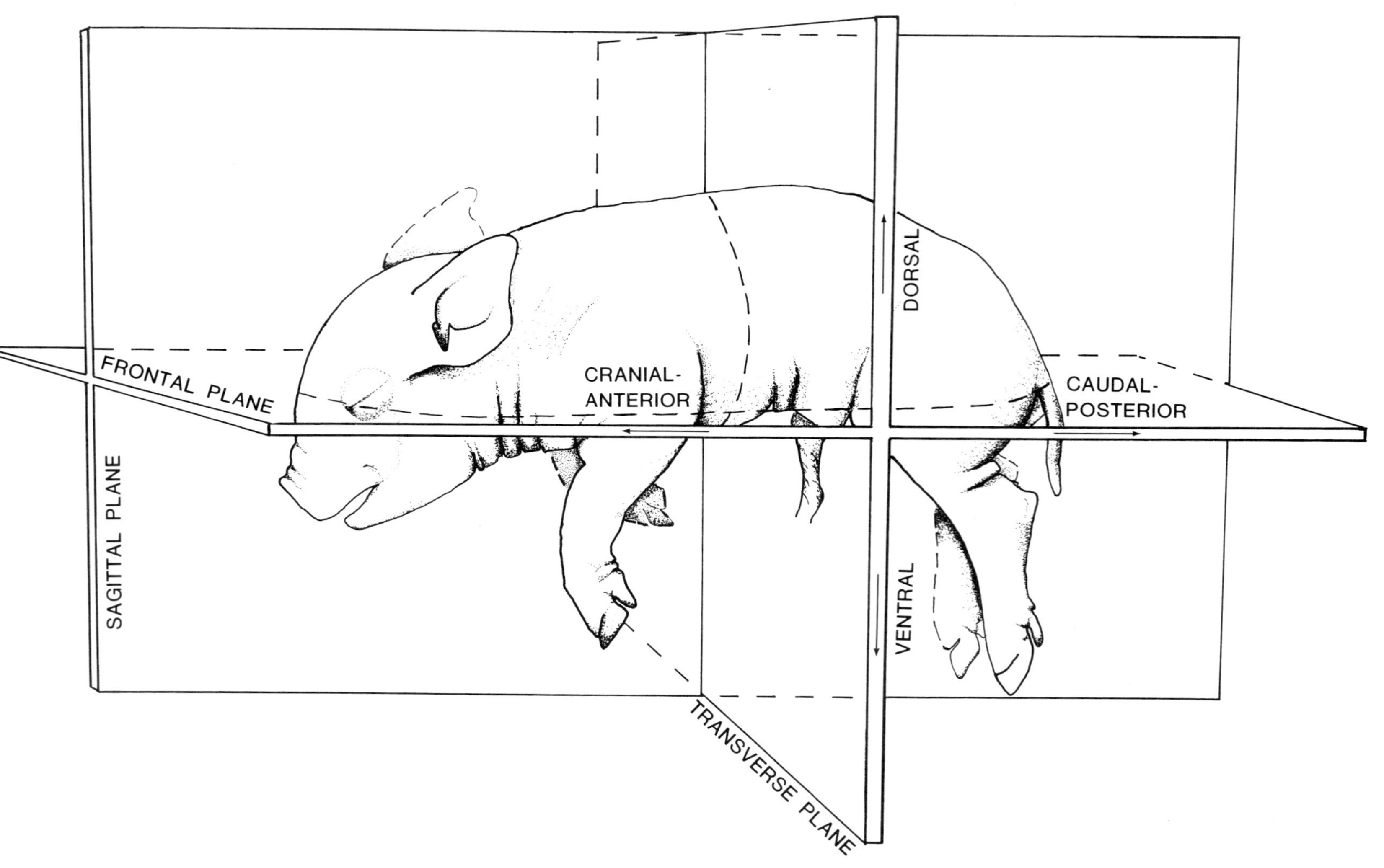

Anatomical planes and directions.

Anatomical Terminology

It is necessary to use a special vocabulary in describing and referring to parts of the mammalian body and their position with respect to each other. The following terms are used throughout the manual with a brief definition for each term. The term "plane" is used to describe the way in which the body could be divided, the term "direction" refers to the orientation of one part of the body to another. The terms "left" and "right" refer to the *pig's* left and right.

Abduction:	Movement of an appendage away from the longitudinal axis of the body.
Adduction:	Movement of an appendage toward the longitudinal axis of the body.
Adipose:	Fat
Anterior:	In front of, toward the head. Example: The head is anterior to the tail.
Bisect:	To cut or divide into two approximately equal parts. Example: Bisecting the fetal pig in the sagittal plane would result in right and left halves.
Caudal:	Toward the tail. Example: The umbilical cord is caudal to the forelegs.
Cephalic:	Pertaining to the head.
Cranial:	Toward the head. Opposite of caudal.
Deciduous:	Shedding. Example: The first set of teeth is deciduous or temporary to be replaced by the permanent set.
Distal:	Farthest from the point of attachment. Example: The wrist is distal to the shoulder.
Dorsal:	The upper surface or back. Example: The back is dorsal to the belly.
Extension:	An action that increases the angle between two parts. Example: The extensor digitorum lateralis muscle extends or straightens out the digits.

Enzyme: A substance that increases and controls the rate of a chemical reaction. Example: The digestion or reduction of a complex carbohydrate to simple sugars.

Fascia: A thin connective tissue layer associated with muscles. Example: One point of origin (attachment) of the external oblique muscle is the lumbodorsal fascia.

Flexion: An action that decreases the angle between two parts. Example: Muscles that pull the head toward the chest.

Frontal: A plane that divides the body into dorsal and ventral parts.

Groin: Junction of the thigh and body.

Homologous: Similarity in structure and origin. Example: The forelimb of the pig and the arm of the human.

Hormone: Endocrine gland secretion that regulates certain bodily activities. Example: Insulin from the pancreas helps to regulate carbohydrate metabolism.

Lateral: Away from the midline, toward the side.

Lumen: A space or cavity within an organ. Example: The lumen of the seminiferous tubules.

Medial: The midline, toward the middle.

Posterior: Toward the tail or rear end. Example: The tail is at the posterior end of the body.

Proximal: Closest to the point of attachment. Example: The shoulder is proximal to the elbow.

Raphe: A seam or line of junction.

Sagittal: A plane dividing the body into right and left parts.

Septum: A partition or wall separating two cavities. Example: The septum pellucidum between the lateral ventricles of the brain.

Sulcus: A groove. Example: The sulci of the cerebrum.

Transverse: A plane at right angles to the longitudinal axis. Also called cross section. This cut would divide the body into anterior and posterior parts.

Ventral: On the underside, opposite of dorsal. Example: The umbilical cord is on the ventral side of the body.

Viscera: Organs of the thoracic and abdominal cavities.

Chapter 1
External Anatomy

(Figures 1.1, 1.2, 1.3, 1.4, 1.5)

BODY

(Figures 1.1, 1.2)

The body is composed of head, neck, trunk, and tail, with the body proper divisible into a *thorax* and an *abdomen*. The cut *umbilical cord* can be seen on the ventral portion of the abdomen near its posterior end. During fetal life, the cord with its contained blood vessels connects the fetus to the placenta on the uterine wall of the female. The external opening of the large intestine, the *anus,* is located immediately under the tail.

HEAD

Located on the head are the *external ears,* the *eyes* with their upper and lower *lids,* and a fairly large *mouth.* The *tongue* with its edging of *papillae* can be seen protruding slightly from the mouth. The *nostrils* or *external nares,* rimmed with tough connective tissue, lie just dorsal to the mouth.

APPENDAGES

The first *digit* or *toe* of both forelimbs and hindlimbs is absent and the second and fifth digits are reduced in size. The only fully functional digits are the third and fourth. The pig is an *ungulate* and its type of locomotion is *unguligrade* in which the weight of the body is borne on the tips of the digits, which are in the form of a *hoof.* In the forelimb, the *wrist* lies just above the digits. The *elbow* can be felt as a bony protuberance on the posterior face of the leg close to the junction of the leg with the body. The *shoulder* is located well above the elbow but is not recognizable by any obvious external feature. Its location, however, can be approximated by feeling for a long bony spine of the *scapula* located about midway between the elbow and the middorsal line of the body.

In the hindlimb, the *ankle* will be seen as a marked protuberance a short distance above the digits. The *knee* is situated on the anterior face of the hindleg on about the same level as the elbow in the foreleg. Dorsal to the knee is the *hip* identified by feeling a bony mass close to the middorsal line.

SEX

(Figure 1.2)

Determine the sex of your specimen. The male is identified by the presence of a swelling, the *scrotal sac,* at the caudal end of the body between the upper ends of the hindlegs and by the presence of the *urogenital opening* just caudal to the umbilical cord. It is also possible to feel the *penis,* a long muscular tubular structure lying under the skin and proceeding caudally from the urogenital opening.

In the female, the urogenital opening is situated immediately ventral to the anal opening. A small fleshy genital papilla projects from the urogenital aperture. In both the male and female, there are two rows of *mammary papillae* or *teats* on the ventral surface of the abdomen with 5 to 7 nipples in each row. After determining the sex of your specimen, examine specimens of the opposite sex.

INTEGUMENT

(Figures 1.3, 1.4, 1.5)

The skin or integument of the mammalian body provides for a variety of functions. It prevents the drying out (dessication) of the underlying structures that it covers, contains pigment cells associated with skin color, regulates temperature through the secretion of the sweat glands and presence of hair, and, through the presence of sensory end-organs such as those for pressure, pain, and tactile sense, keeps the body in contact with the external environment.

The two well-defined layers of the skin that can be seen in a microscopic preparation are the upper *epidermis* and lower *dermis*. The epidermis is the thin irregular outermost border composed of an external layer, the *stratum corneum,* and an internal layer, the *stratum germinativum*. The stratum corneum consists of dead, somewhat flattened cells, the stratum germinativum of living cells that are continually dividing to renew the corneal layer, which gradually wears away.

Lying below the epidermis is a much thicker layer, the *dermis* or *corium*. Within this region are sweat glands, *sebaceous* (oil) glands, hair follicles, certain sensory end-organs, smooth muscles associated with hair, and fat cells. Sweat glands are long coiled structures with ducts leading upward through the dermis and epidermis to open on the surface of the skin. Sebaceous glands secrete oil, occur as small groups of cells, and empty into the hair follicles. Each hair is composed of a *hair shaft* which arises from a *root* deep in the dermis, both enclosed in a *hair follicle*. The shaft of the hair extends above the surface of the skin and is constantly renewed from the hair root. Small smooth muscles, *arrectores pilorum,* are associated with the hair and by their contraction tend to raise the hair.

A broad band lying below the dermis is the *subcutaneous* layer. This is composed of connective tissue cells and fibers. This layer is called *fascia* where it is associated with striated muscle.

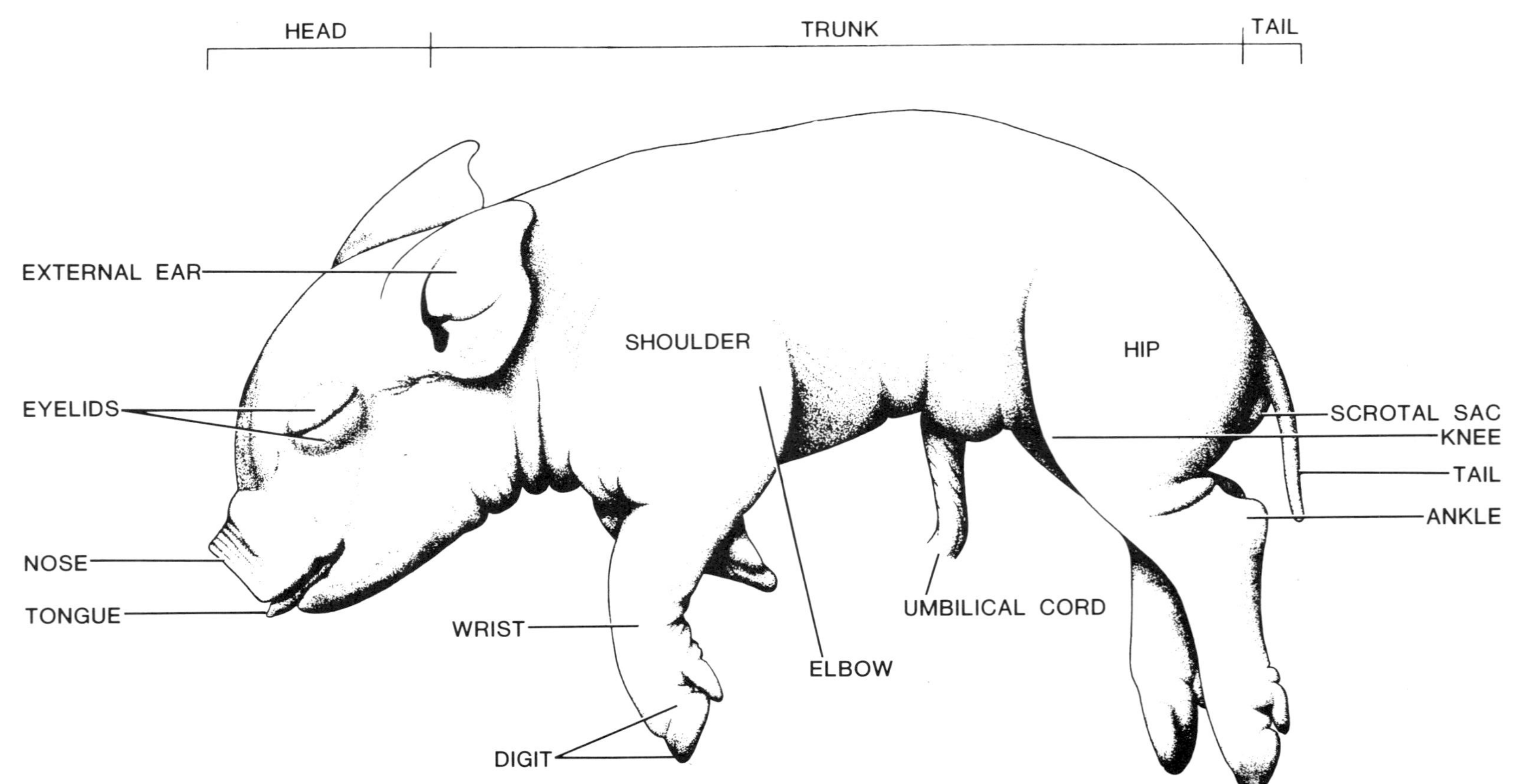

FIGURE 1.1. **External features.**

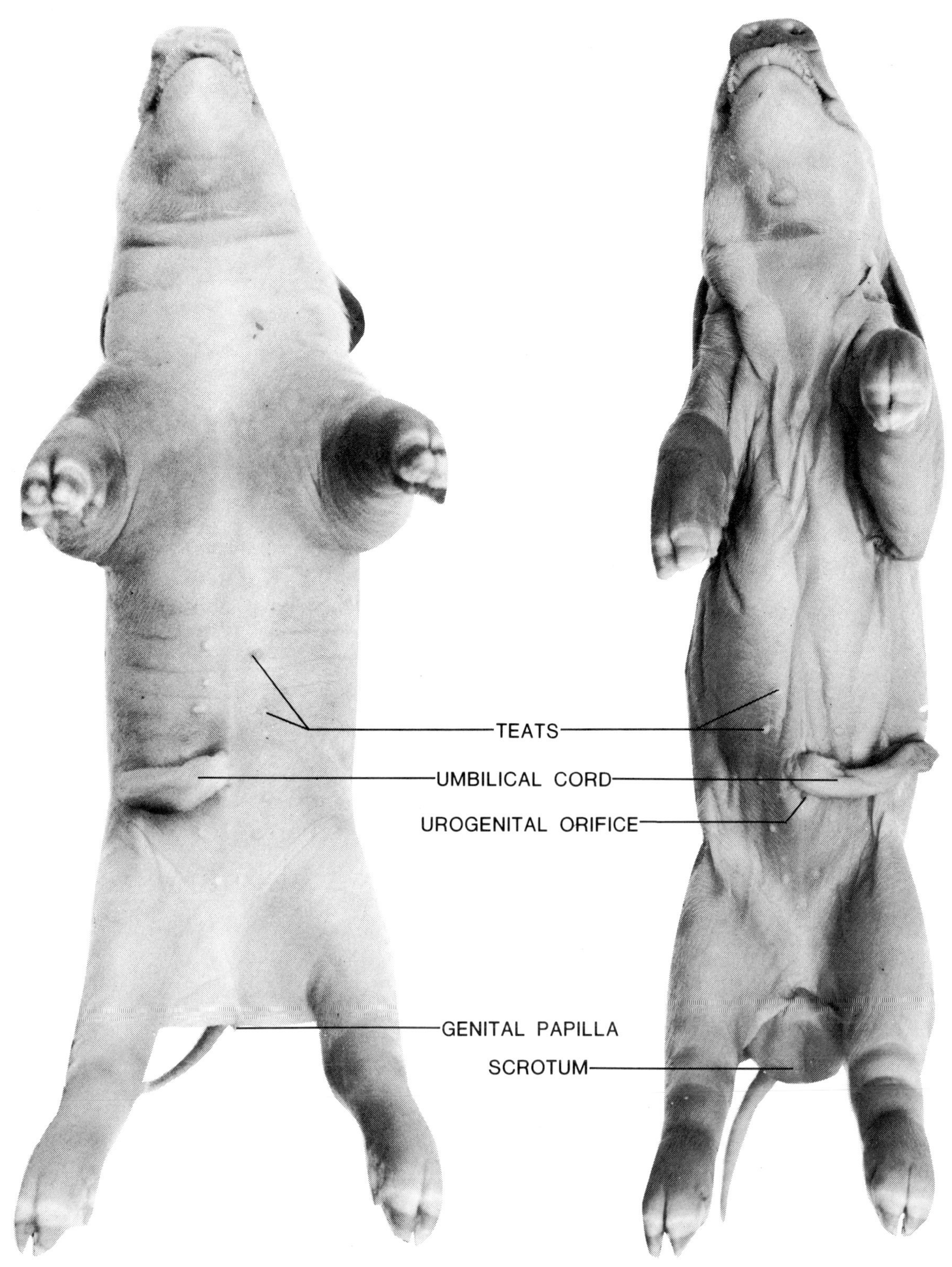

FIGURE 1.2. Ventral view of female (left) and male (right).

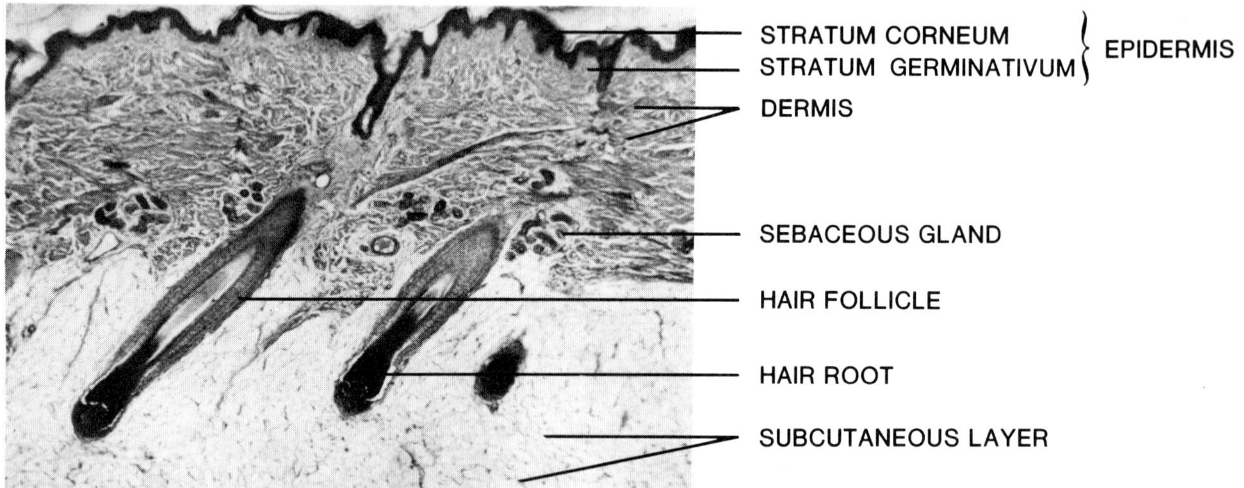

FIGURE 1.3. Vertical section through skin (photomicrograph ×100).

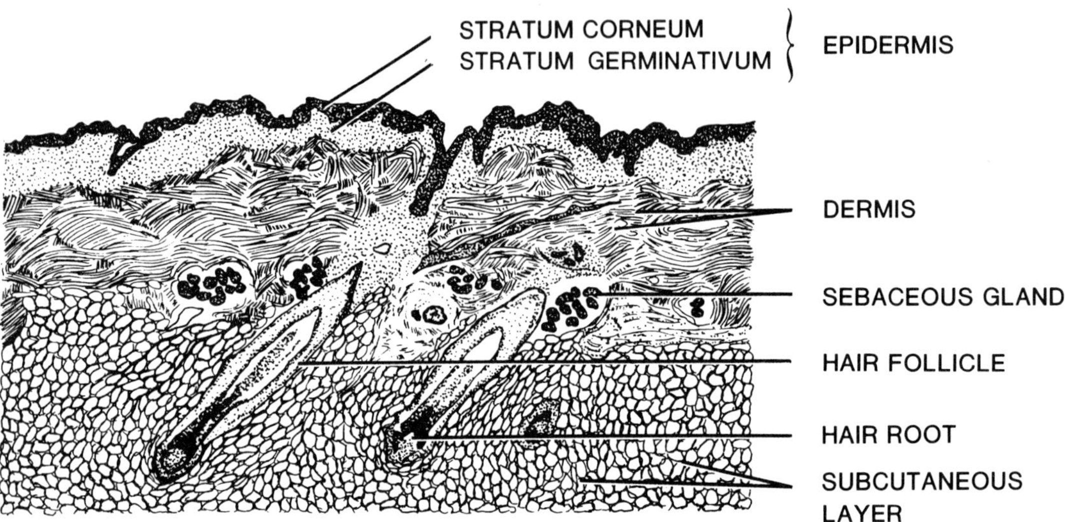

FIGURE 1.4. Vertical section through skin (diagrammatic).

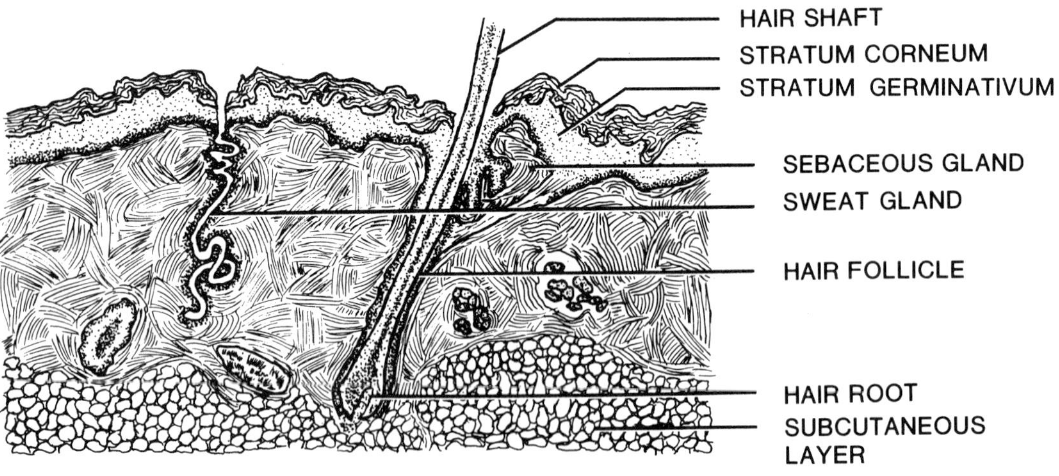

FIGURE 1.5. Vertical section through skin (diagrammatic).

Chapter 2
Skeletal System

(Figures 2.1, 2.2, 2.3, 2.4, 2.5, 2.6, 2.7, 2.8, 2.9, 2.10)

CHAPTER OBJECTIVES

To identify the parts of the skeletal system.

INTRODUCTION

The two major divisions of the skeletal system are the axial and appendicular. The axial skeleton consists of the skull, vertebrae, ribs, and sternum. The appendicular skeleton consists of the pectoral and pelvic girdles and the appendages associated with them.

The cat skeleton will be studied in the laboratory but drawings of both fetal and adult skeletons of the pig are included here for comparison. The material in this chapter, both text and figures, is an abbreviation of the chapter on the skeletal system by Robert B. Chiasson and Ernest S. Booth, *Laboratory Anatomy of the Cat,* 6th Edition, 1977, Wm. C. Brown Publishers, Dubuque, Iowa.

Study both mounted and disarticulated skeletons. The mounted skeleton will show the relation of each bone to the other bones. Since most of the time will be spent with the disarticulated skeleton, however, it will be necessary to learn all the bones of the body.

The skeleton can be divided into two parts: the axial skeleton (bones of the skull, spinal column, and thorax) and the appendicular skeleton (bones of the thoracic and pelvic girdles and appendages). Spend some time studying the entire mounted skeleton, becoming familiar with the larger bones and their relationship to each other. Spend the rest of your time studying the disarticulated skeleton.

AXIAL SKELETON

I. Skull (figs. 2.4, 2.5, 2.6, 2.7).
  A. Cranium, caudal skull bones enclosing the brain.
    1. Frontal, paired bones with processes at the caudal border of the orbit directed toward the zygomatic arch.
    2. Parietal, paired bones. The intersection between the frontal and parietal bones is called the *bregma.*
    3. Interparietal, a small triangular bone between the two parietals and the supraoccipital.
    4. Temporals are paired bones, and each is divisible into three parts.
      a. Squamosal.
      b. Periotic *or* mastoid *or* petromastoid.
      c. Tympanic bulla.
    5. Occipital is the most caudal skull bone and is fused from four parts.
      a. Supraoccipital, the caudal wall of the cranium.
      b. Basioccipital, the caudal floor of the cranium.
      c. Exoccipitals, paired bones with *condyles* for articulation with the atlas.
    6. Basisphenoid has two wings extending into the orbit and forms the caudal pterygoid and hamular processes.
    7. Presphenoid, lateral portions are called *orbitosphenoids.*
    8. Ethmoid has a caudal perforated *cribiform plate* at the most rostral end of the cranial cavity. The remainder of this bone is a part of the splanchnocranium.
  B. Facial Bones. These are bones of the splanchnocranium surrounding the digestive and respiratory systems and the special sense organs.
    1. Nasals, paired bones roofing the ethmoid.
    2. Incisive bones, paired bones containing the incisor teeth.
    3. Maxilla, all of the upper teeth except the incisors are socketed in these paired bones.
    4. Lacrimals, small paired bones at the rostral border of the orbit. The lacrimal canal passes through the dorsal rostral corner of this bone.
    5. Zygoma is the central bone of the zygomatic arch. Connects the maxilla and squamosal and has a frontal process.
    6. Vomer is unpaired, dorsal to the palatine bones and supports the ethmoid.
    7. Ethmoid (also see no. 8 under cranium), consists of a central perpendicular plate set in the vomer (see no. 6) and thin, turbinate bones on the sides of the perpendicular plate. The caudal perforated surface is called the cribiform plate.
    8. Palatines, paired bones forming the rostral roof of the mouth (the hard palate).
    9. Mandible, a paired bone that articulates with its fellow in the rostral midline by the *mandibular symphysis.* They are often fused at this point in old specimens. All of the lower teeth are socketed in the mandible. Caudally, there is a dorsal *coronoid* process for attachment of the temporalis muscle, a ventral *angular* process for attachment (medially) of the pterygoid muscles, and between these processes, a *condyloid* process

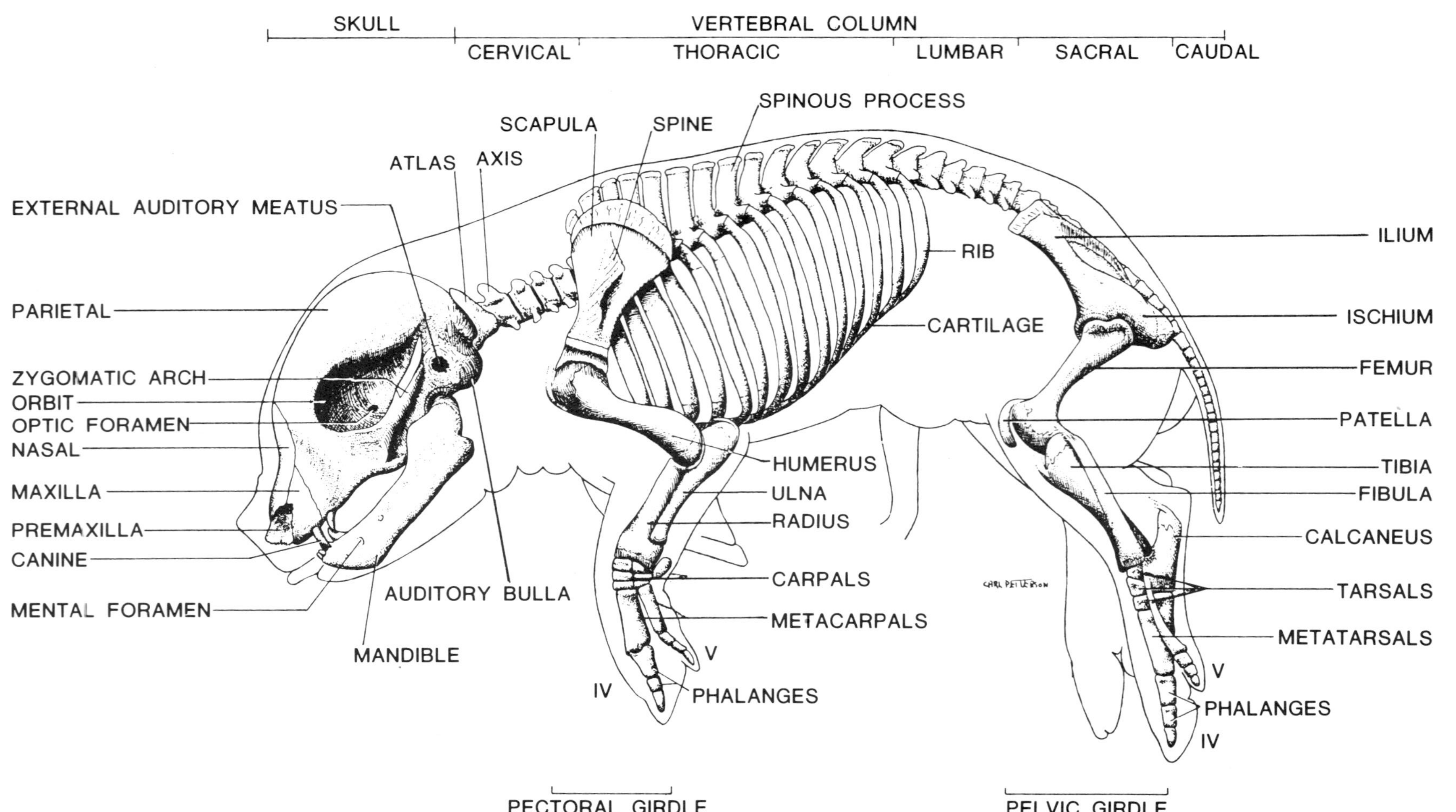

FIGURE 2.1. Skeleton of fetal pig, lateral view.

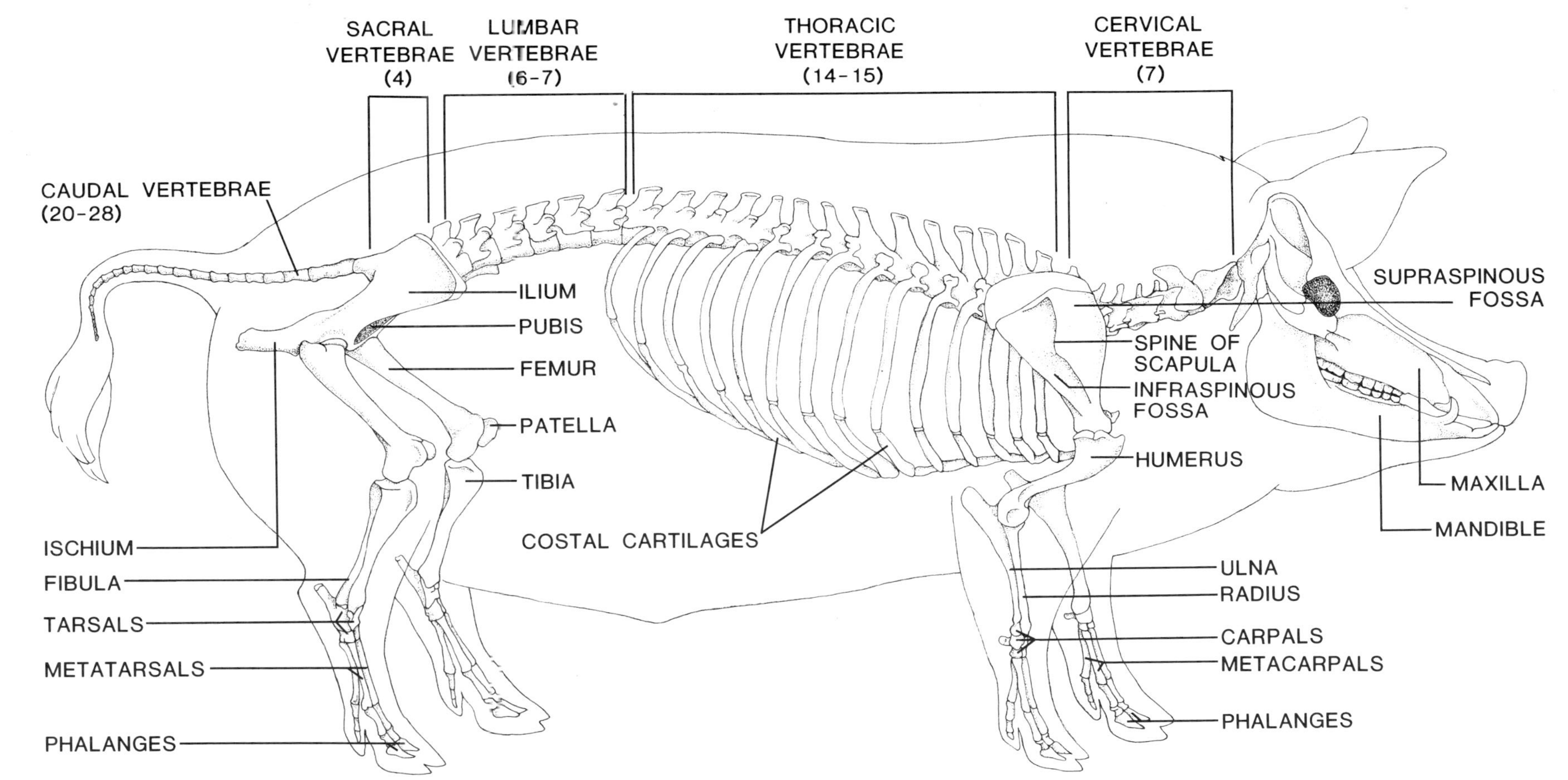

FIGURE 2.2. **Skeleton of adult pig, lateral view.**
(Modified from Sisson and Grossman, *Anatomy of the Domestic Animals,* 5th ed. Philadelphia, Pa.: W. B. Saunders Company, © 1975.)

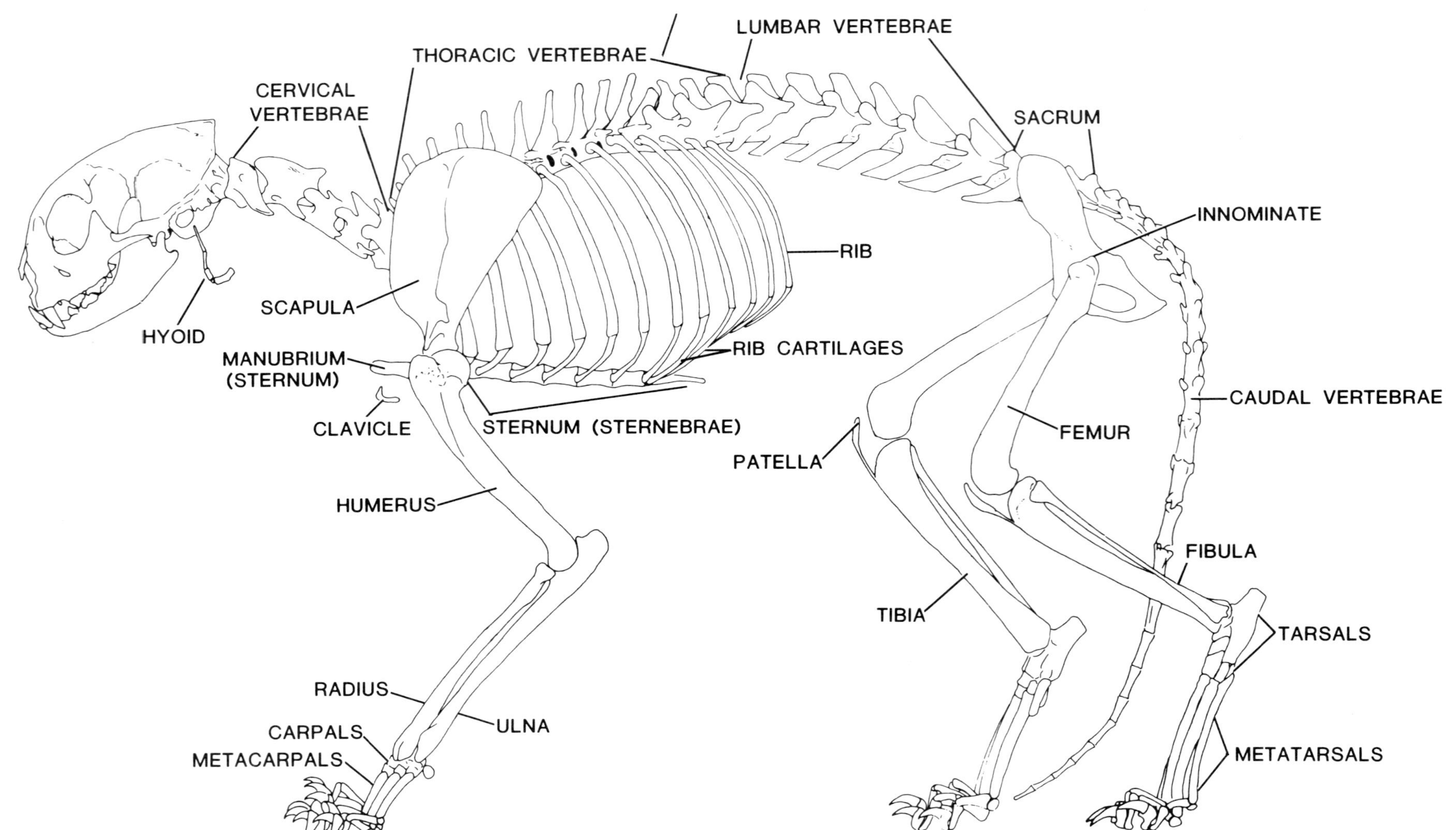

FIGURE 2.3. Skeleton of cat, lateral view.

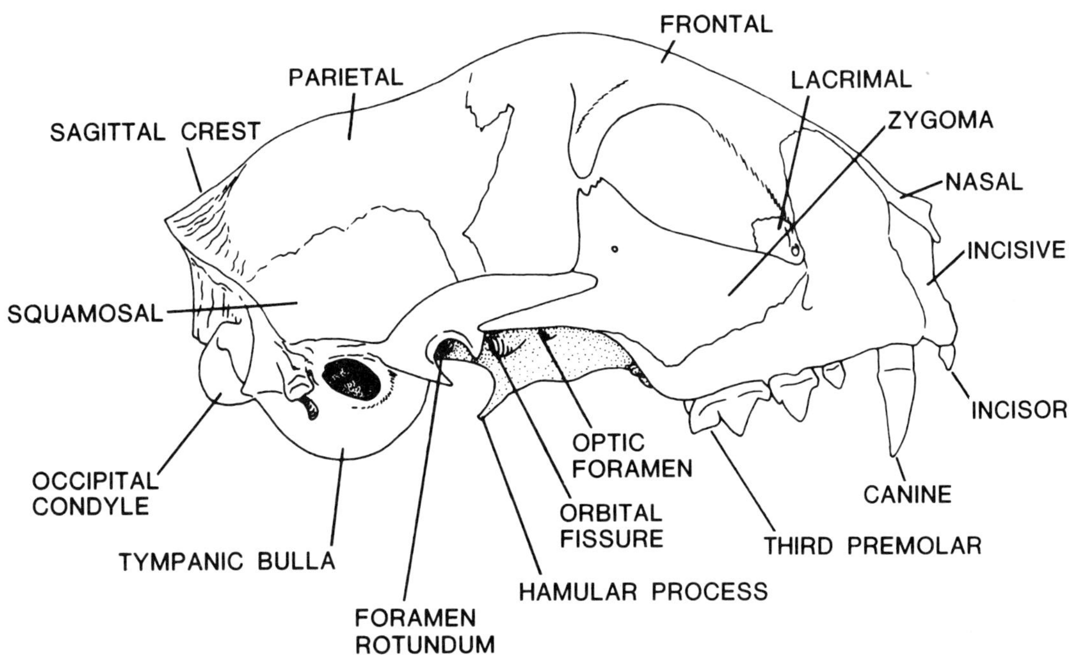

FIGURE 2.4.   Lateral view of skull.

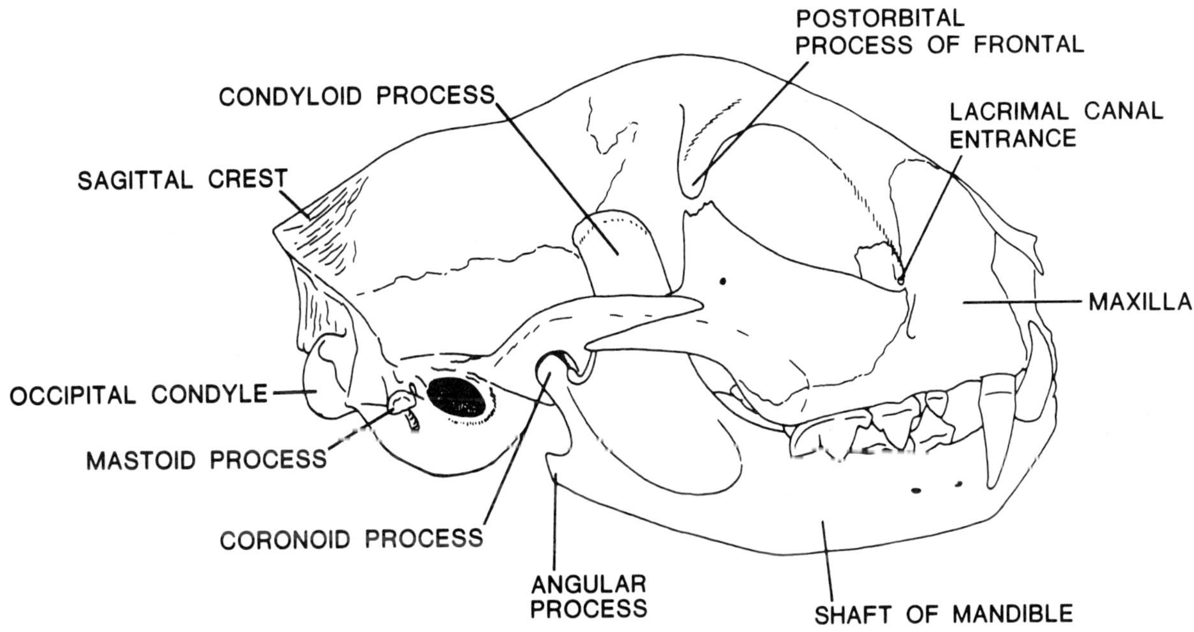

FIGURE 2.5.   Skull and mandible, lateral view.

for articulation with the squamosal. A single, lower jawbone (on each side) is a primary mammalian character.

10. Hyoid, consists of eleven bones, ten paired units and an unpaired *basihyal*. The cranial horn has four pairs of bones: a *ceratohyal* (articulating with the *basihyal*), *epihyals, stylohyals,* and *tympanohyals* (attached to the tympanic bulla). The caudal horn consists of a single pair of *thyrohyals* between the basihyal and the thyroid cartilage of the larynx.

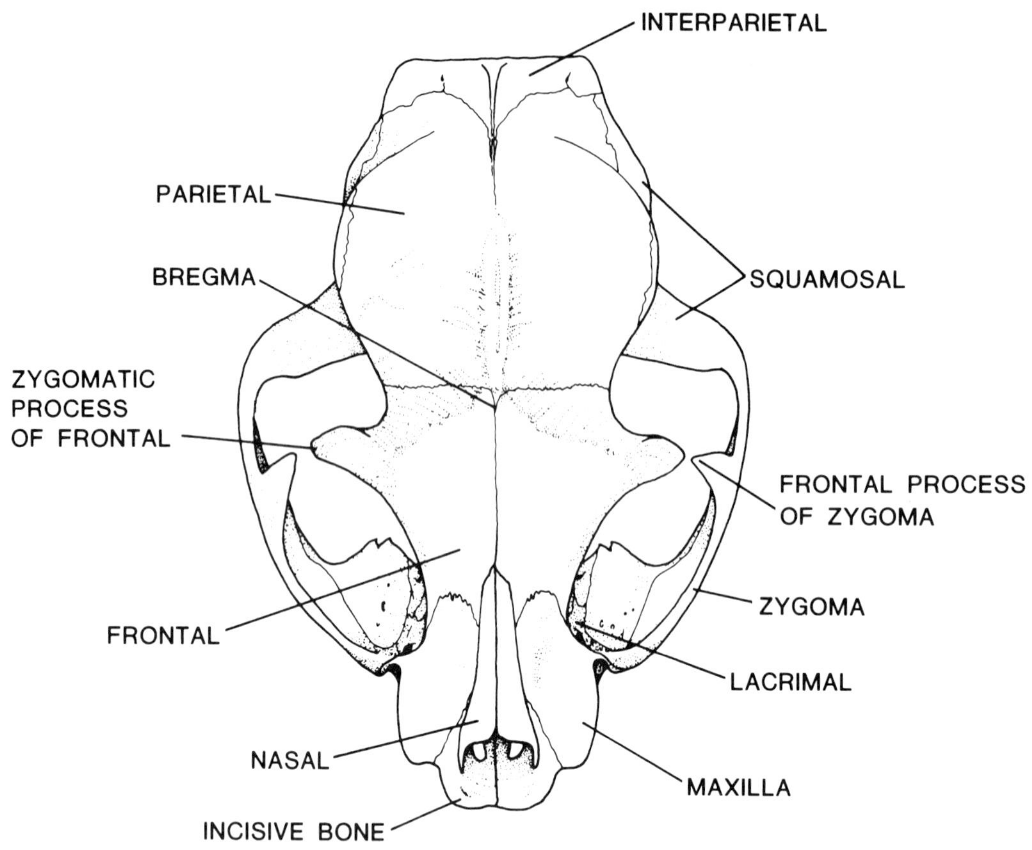

FIGURE 2.6. Skull, dorsal view.

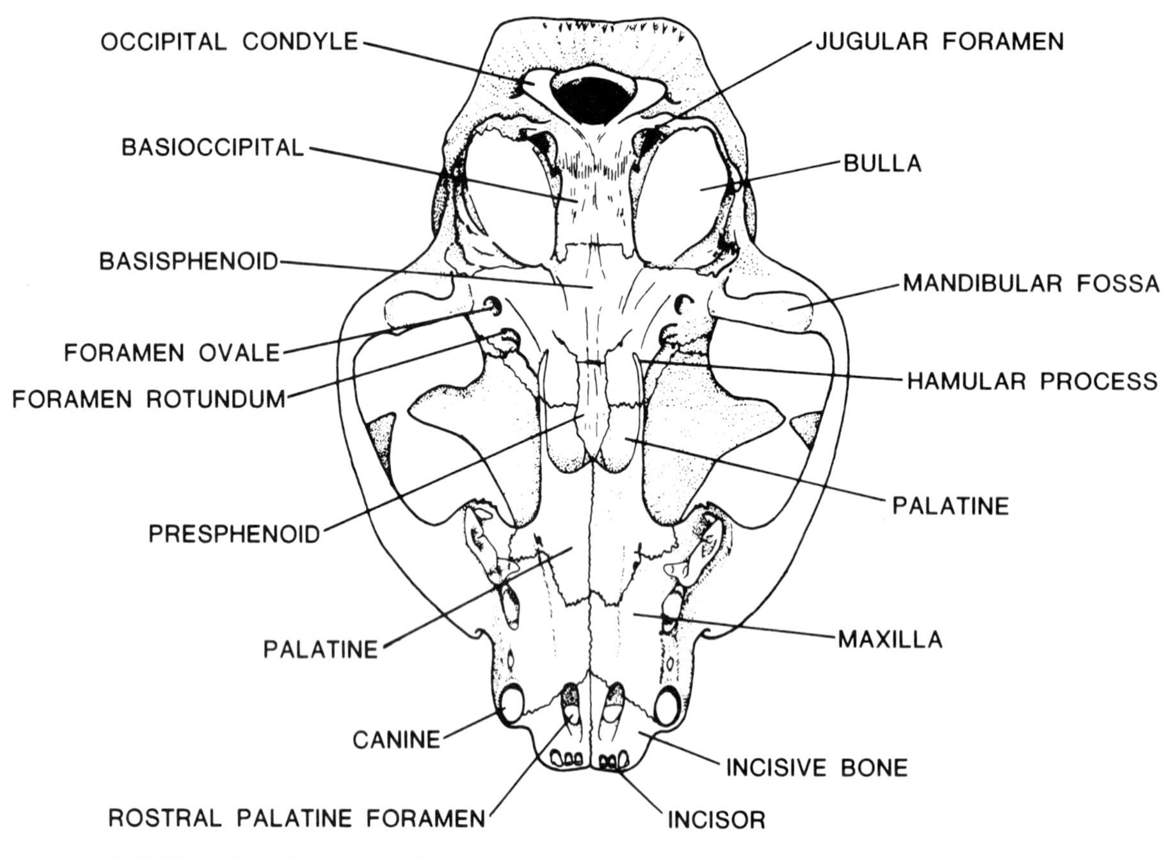

FIGURE 2.7. Skull, ventral view.

II. Vertebrae, Ribs, and Sternum (figs. 2.3, 2.8).
  A. Vertebral Column.
    1. Cervical Vertebrae. There are seven cervical vertebrae in mammals. The first is called the *atlas;* the second, the *axis* (see fig. 2.8). Each cervical vertebra, except the seventh, can be recognized by the transverse foramen, a small hole on either side of the vertebral canal for the passage of the vertebral artery.
    2. Thoracic Vertebrae. There are thirteen of these. They may be identified by their long dorsal spine (see fig. 2.8), and by the fact that ribs are attached to them.

       Nine of the thirteen pairs of ribs are known as *true ribs,* for they are attached to the sternum by *costal cartilages.* The next three pairs of ribs are joined to the ninth rib rather than to the sternum. These are called *false ribs.* The thirteenth pair of ribs does not join the sternum nor the other ribs. They are called *floating ribs.*

       The rib (figs. 2.3, 2.8) is composed of a *vertebral portion* (which is ossified or bony), and a *cartilaginous portion.* The vertebral portion is made up of the *head* or *capitulum* (articulating with the centum of the vertebra), *tuberculum* (articulating with the transverse process), neck (narrow part between the capitulum and the tuberculum), and the *shaft* (longer portion from the tuberculum to the ventral end of the rib). There are neither necks nor tuberculae on the last three pairs of ribs.

       The sternum is composed of eight segments called *sternebrae.* The cranial sternebrum is the *manubrium,* and the caudal one is the *xiphoid process* or the *xiphisternum.*
    3. Lumbar Vertebrae. There are seven of these. They may be readily identified by the large transverse process that projects ventrally and cranially (see fig. 2.8).
    4. Sacral Vertebrae or Sacrum. This is a single bone in adults formed by the union of at least three sacral vertebrae.
    5. Caudal Vertebrae. The number of these varies a great deal, depending on the length of the tail, but there are usually at least twenty. They may be confused with the bones of the sternum and the phalanges unless one is careful to note the exact shape of each of these three types of bones. Note the "dumbbell" shape of each caudal vertebra, caused by the wide projecting processes at each end of the bone.
    6. Typical Vertebra (see fig. 2.8). On one of the lumbar vertebrae locate the *centrum* (large, ventral body portion), *neural canal* or *vertebral foramen* (through which the spinal cord passes), *neural spine* or *spinous process* (extending dorsally from the neural canal), *transverse processes* (lateral-ventral outgrowths at the base of the neural canal), and the *intervertebral foramina* (narrowed part of the neural arch near the point of attachment of the centrum), where the spinal nerves branch out from the spinal cord.

APPENDICULAR SKELETON

I. Pectoral Girdle and Appendage (figs. 2.3, 2.9).
  A. Pectoral Girdle.
    1. Scapula, paired "shoulder blades." In the cat these bones are the only *functional* part of the pectoral girdle. A *spine* divides the lateral surface into cranial (*supraspinous*) and caudal (*infraspinous*) *fossa.* A caudal projecting process from the spine is the *metacromion* process, and the ventral continuation of the spine is the *acromion* process. Opposite the acromion process, on the medial surface of the scapula, is the hook-shaped *coracoid* process. Between the acromion and coracoid processes is the *glenoid* fossa for articulation with the head of the humerus.
    2. Clavicles are small, paired bones buried in muscles, with no articulation.
  B. Pectoral Appendage.
    1. Humerus. The proximal end of the humerus is characterized by a *head* that articulates with the scapula and *greater* and *lesser tuberosities* that serve for muscle insertions. The distal end is characterized by a two-planed articular surface: the rounded lateral half, the *capitulum,* articulates with the radius and the concave lateral surface, the *trochlea,* articulates with the ulnar *semilunar notch.*
    2. Radius is the lateral bone of the forearm.
    3. Ulna is the medial bone of the forearm. The proximal end of the ulna has a *semilunar notch* that articulates with the humerus and a

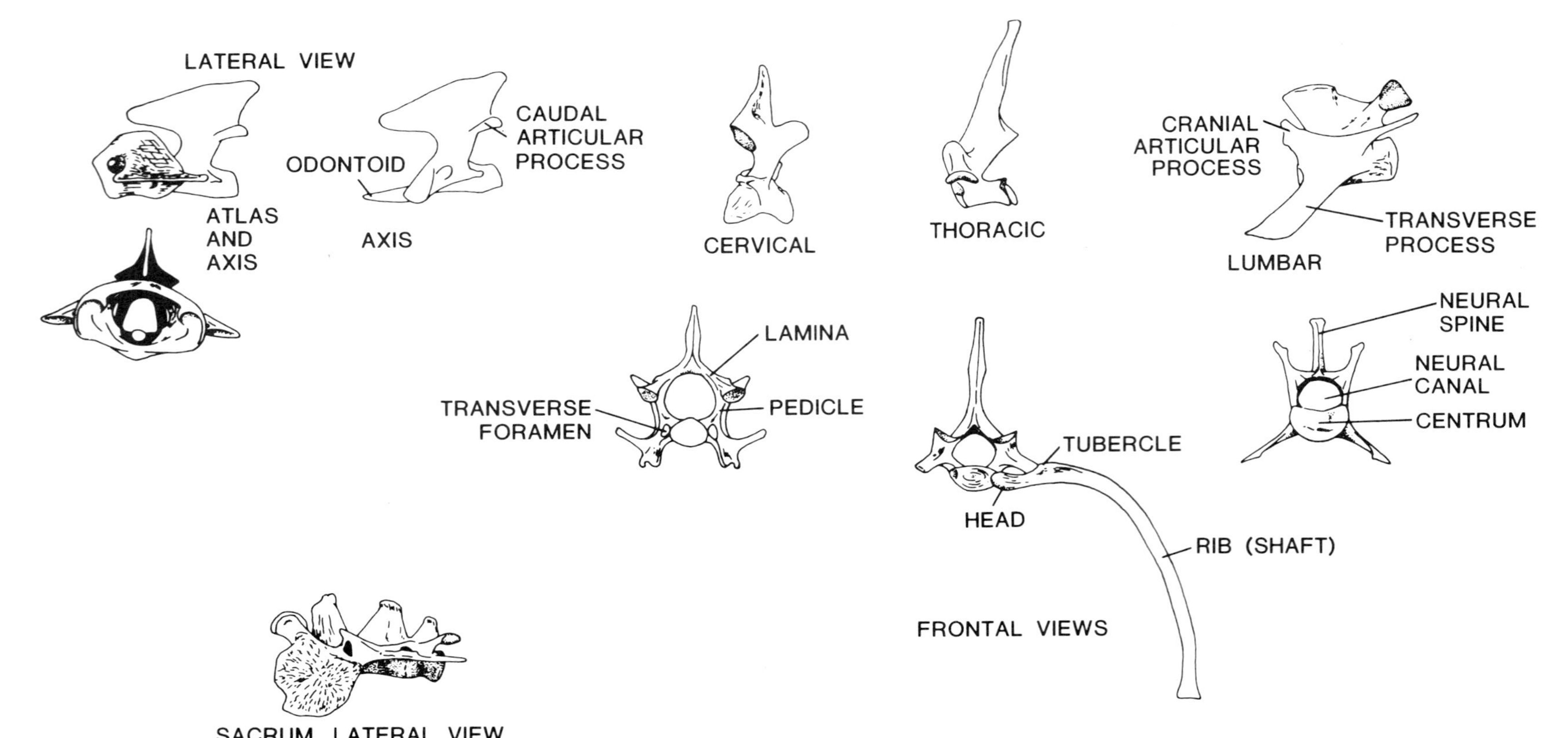

FIGURE 2.8. Typical vertebrae, lateral (top) and frontal (lower) views.

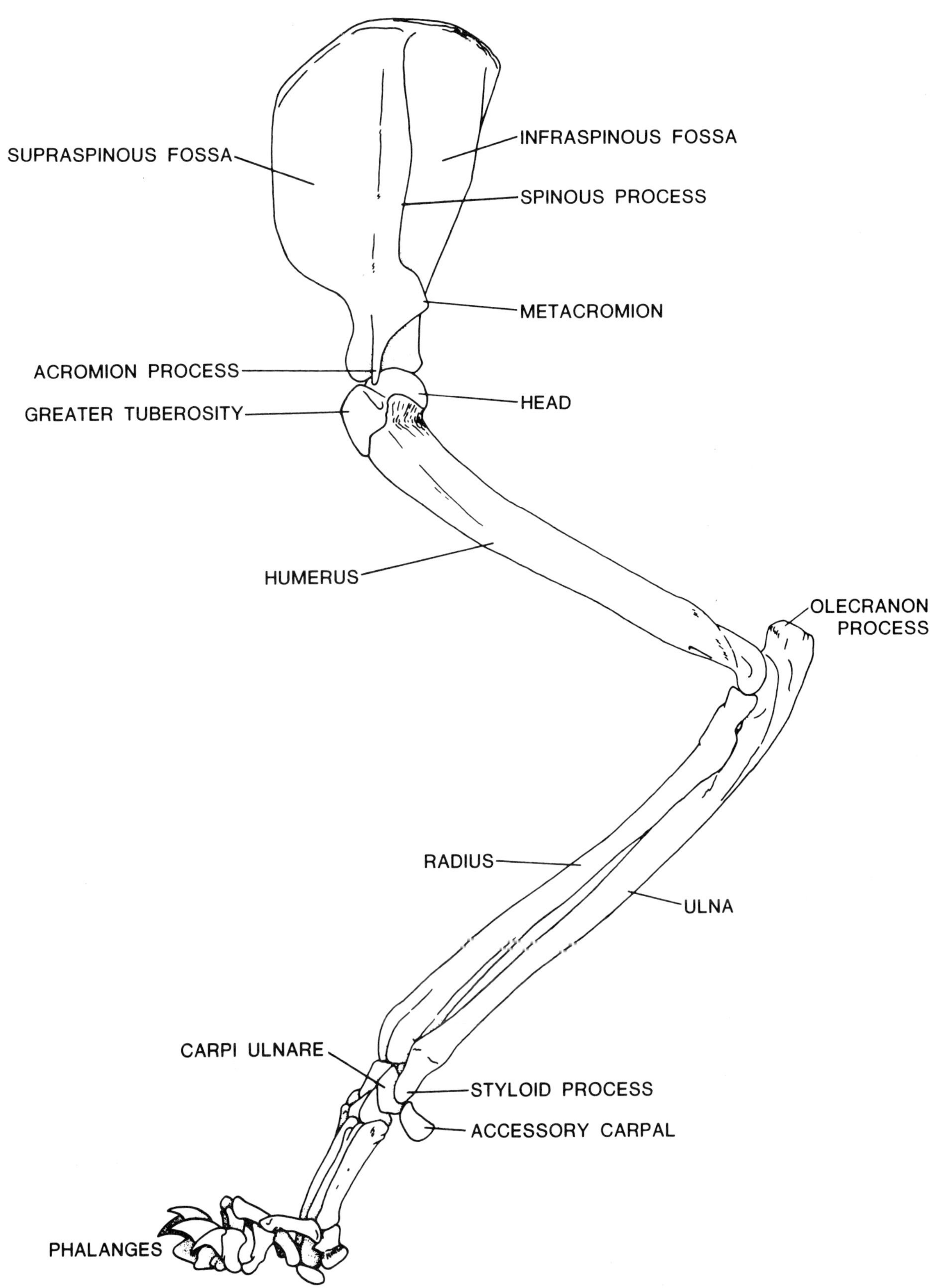

FIGURE 2.9. Pectoral girdle and limb, lateral view.

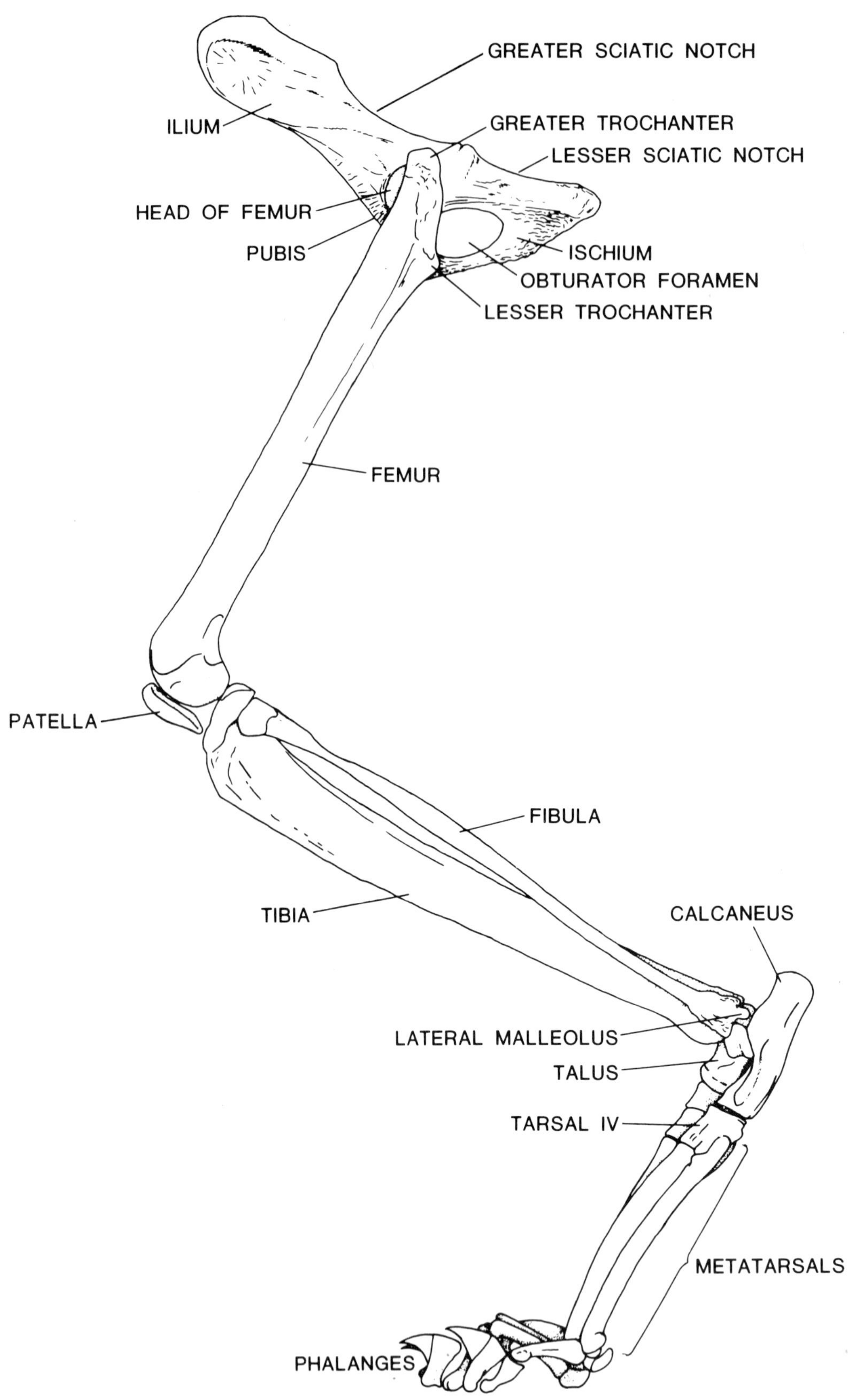

FIGURE 2.10. Pelvic girdle and limb, lateral view.

projection beyond the semilunar notch called
the *olecranon process.*
        4. Carpal bones.
        5. Metacarpals are the bones of the paws. There
           are five metacarpals in each manus, one for
           each digit.
        6. Phalanges are the bones of the digits. The
           first digit (pollex) has only two phalanges but
           each of the other four digits have three
           phalanges. All five digits of each hand have
           terminal horny claws on the distal phalanges.
II. Pelvic Girdle and Appendage (figs. 2.3, 2.10).
    A. Pelvic Girdle.
        1. Innominate bone is composed of four separate
           units that fuse together to form each
           innominate. The two innominates are joined
           in the ventral midline by the *pubic symphysis*
           and articulate dorsomedially with the sacrum
           by the *iliosacral* joint.
           a. Ilium is the cranial-most portion of the
              innominate.
           b. Ischium is the dorsocaudal portion of the
              innominate.
           c. Pubis is the ventrocranial portion of the
              innominate.
           d. Acetabular bone (condyloid bone) is a
              small triangular bone within the
              acetabulum.
           e. Acetabulum is the socket for articulation
              with the head of the femur.
           f. Obturator foramen is a large opening in
              the caudal half of the innominate, between
              the ischium, pubis, and acetabulum.
        2. Femur is the large bone of the thigh. The
           proximal end has a head that articulates with
           the acetabulum and *greater* and *lesser
           trochanters* for muscle attachments. The
           head of the femur is set at an angle of about
           45° to the *shaft* of the femur. The distal end
           has two articular *condyles* separated by an
           intercondyloid fossa.
        3. Tibia is the large medial bone of the leg. The
           proximal end has two articular condyles
           separated by a spine. The distal end has the
           medial malleolus.
        4. Patella is a small sesamoid bone cranial to
           knee joint, the "knee cap."
        5. Fibula is the slender, splintlike lateral bone of
           the leg.
        6. Tarsals are the ankle bones.
        7. Metatarsals are the bones of the foot or *pes*.
           The cat has five metatarsals, one for each
           digit and a rudimentary first metatarsal just
           distal to the external cuneiform.
        8. Phalanges are the bones of the digits. The cat
           has only four digits on the foot, and there are
           three phalanges in each digit.

Chapter 3
Muscular System

(Figures 3.1, 3.2, 3.3, 3.4, 3.5, 3.6, 3.7, 3.8)

CHAPTER OBJECTIVES

1. To identify the gross muscle groups and their relationship to other organs of the body.
2. To identify specific muscle types within each group.
3. To examine the microscopic structure of each muscle group.

INTRODUCTION

Muscles are classified as to structure and function. The largest mass of muscle tissue is called *skeletal* because these muscles are attached to the bony framework; *voluntary* because the muscles are under the control of the individual organism; and *striated* because of the alternating light and dark bands as seen in microscopic section.

Muscles that are associated with the viscera are known as *smooth* because they lack the striations characteristic of skeletal muscle. *Cardiac* muscle is limited to the heart.

The muscles of the fetal pig are not commonly used as representative of the system since they are not fully developed and many are easily torn. With reasonable care, however, a satisfactory dissection can be made and muscle attachments, relationships, and functions determined.

The three parts of a typical muscle are the *origin,* the end attached to a more or less rigid part of the skeleton; the *insertion,* the end attached to some part of the skeleton that moves when the muscle contracts; and the *belly,* the thickened middle portion between the two points of attachment.

For all the muscles you dissect, proceed by first carefully determining the area covered by the muscle, then completely free it between the two points of attachment as far as is possible. When it becomes necessary to expose deep muscles, cut the overlying muscle through the *belly* and reflect the two ends.

Muscles are attached to various parts of the skeletal system, especially at the point of insertion, by means of a compact group of fibrous connective tissue cords called a *tendon.* For a number of muscles a broad, flat sheet of connective tissue called an *aponeurosis* performs the same function as the tendon.

DIRECTIONS FOR THE DISSECTION OF THE MUSCLES

Using a sharp scalpel, start the incision on the ventral side at the base of the throat, making sure you are cutting *only* through the skin. Continue the incision caudally to the umbilical cord, then around the cord on the right side (the specimen's right side) and from there to the level of the hindlegs. From the midventral incision, carry the cut down the medial surface of the right hindleg and then do the same for the skin of the right foreleg.

Grasp the cut edge of the skin with a forceps and begin peeling it away from the underlying tissue. In the *thoracic* or chest region, muscles will be seen as soon as the skin is freed. Sometimes there is a tendency for the muscles to remain attached to the skin. If this occurs, grasp the cut edge of the skin with the fingers of one hand while pushing on the muscle with the thumb of the other hand.

In the lower thoracic area and in the abdominal region, the muscles are covered with a thin layer of *adipose* (fat) tissue, fascia, and muscle fibers. The latter constitutes the superficial muscle called *cutaneous maximus.* This entire layer must be removed by carefully picking it away with the forceps until a definite layer of muscle fibers is revealed. This is the *external oblique.* Note how the fibers of this muscle are directed ventrocaudally.

As you work with the muscles, they can be seen more easily if the moisture on them is sponged away with paper toweling.

MUSCLES OF THE BACK, NECK, AND SHOULDER
(Figures 3.1, 3.2)

1. *Latissimus dorsi.* A broad muscle directed downward and cranially around the sides of the thoracic region. If the muscle fibers are not immediately apparent, carefully pick away the surface layer from the sides of the chest until the fibers come into view. Origin, from the lumbar and some of the last thoracic vertebrae and from the *lumbodorsal fascia;* insertion, by means of a tendon into the proximal end of the humerus on its medial face; action, moves the forelimb dorsally and caudally.
2. *Trapezius.* A broad muscle cranial to the preceding and covering part of its craniodorsal edge. Origin, from the occipital bone of the skull and from the spinous processes of the first ten thoracic vertebrae; insertion, on the spine of the scapula; action, draws the scapula medially.
3. *Brachiocephalic.* Carefully remove the mass of cutaneous muscle and gelatinous connective tissue from the side of the neck and observe the broad flat strap of muscle extending obliquely from the back of the skull to the foreleg. Free and raise the caudal edge of the parotid gland and note that the muscle consists of two parts, the dorsal portion called the *cleido-occipitalis,* and the ventral

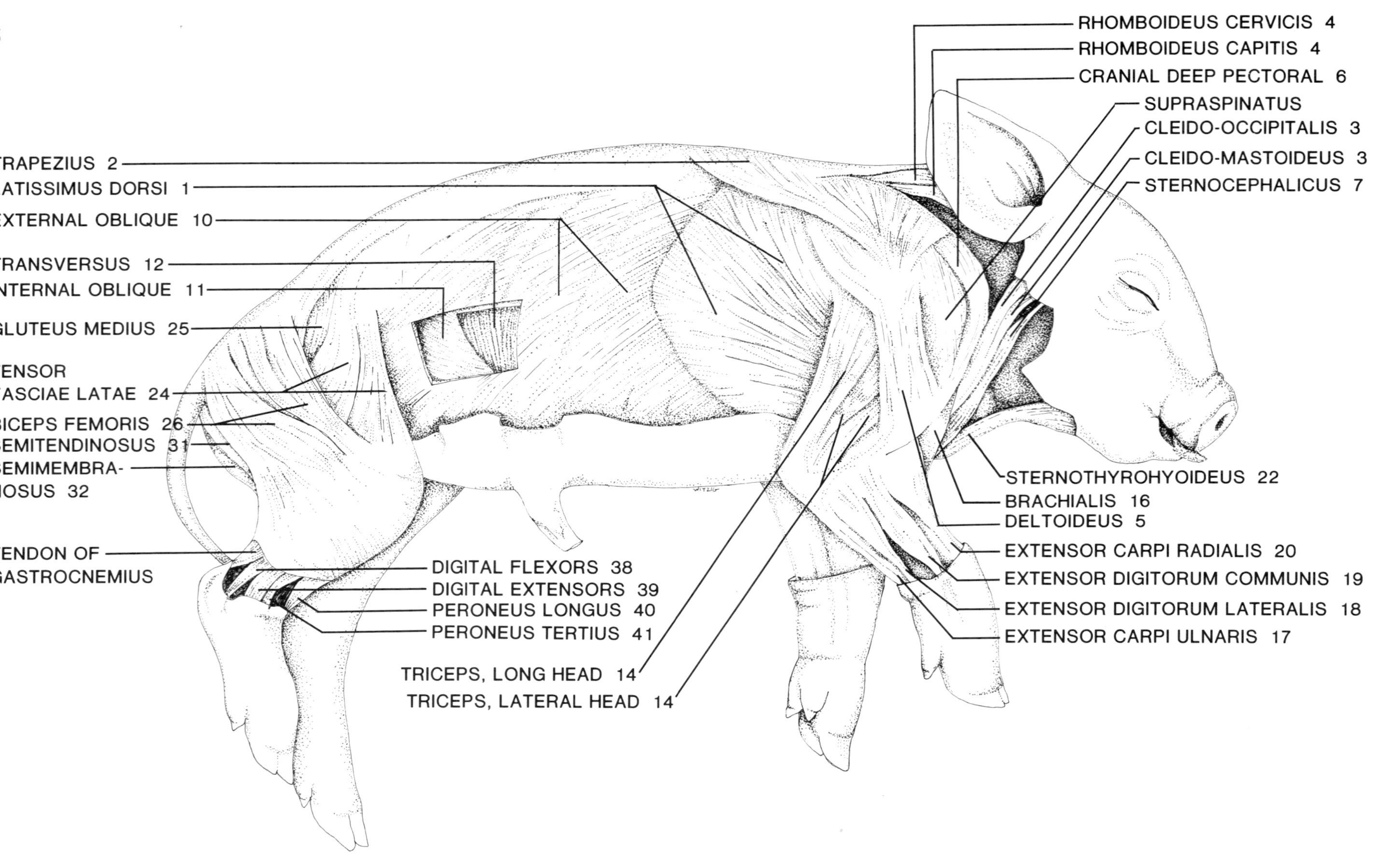

FIGURE 3.1. Muscles, lateral view.

portion called the *cleido-mastoideus*. The two parts unite near the level of the lower end of the deltoid muscle. Origin, lambdoidal ridge of the occipital bone and from the mastoid process; insertion, distal end of the humerus; action, flexes foreleg.

4. *Rhomboideus.* Pull the forelegs together under the body, bisect the trapezius through the belly and reflect the cut ends. This will expose the rhomboideus which consists of three parts. The *rhomboideus thoracis* is a broad band of muscle fibers originating on the last few cervical and first six thoracic vertebrae and inserting on the vertebral border of the scapula; action, draws scapula forward. The *rhomboideus cervicis* is a long slender muscle originating on the last few cervical and first six thoracic vertebrae and inserting on the vertebral border of the scapula; action, draws scapula forward and medially. The *rhomboideus capitis* is also a long slender muscle parallel with the former muscle but originating on the occipital bone and inserting on the vertebral border of the scapula; action, draws scapula forward.

5. *Deltoideus.* A fairly broad muscle lying between the triceps and the supraspinatus. Origin, from the spine of the scapula; insertion, on the deltoid ridge of the humerus; action, flexes the humerus.

6. *Cranial deep pectoral.* A slender muscle lying along the cranial border of the supraspinatus. Origin, from the sternum at its cranial end; insertion, into the aponeurosis at the dorsal end of the supraspinatus; action, adduction of forelimb.

7. *Sternocephalicus.* A long muscle lying ventral to the brachiocephalic. Origin, from the cranial end of the sternum; insertion, by means of a long tendon into the mastoid process. The tendon is readily seen by removing the parotid gland; action, flexes the head.

8. *Splenius.* Bisect the rhomboideus cervicis and the rhomboideus capitis to expose the thick muscle mass which lies ventral to them. Origin, from a cervical ligament on the middorsal line of the neck; insertion, on the dorsal portion of the occipital bone; action, raises the head.

MUSCLES OF THE CHEST AND ABDOMEN
(Figures 3.1, 3.3)

9. *Pectoralis major.* A broad fan-shaped muscle. Origin, from the sternum; insertion, into the proximal end of the humerus; action, draws the forelimb toward the chest (*adduction*).

10. *External oblique.* This muscle, together with the next two, comprise the musculature of the lateral abdominal wall. Observe that the fibers of the external oblique are directed ventrocaudally. Origin, by slips from the caudal ribs and from the lumbodorsal fascia; insertion, by means of an aponeurosis into the *linea alba* (a ventral longitudinal ligament); action, constricts abdomen.

11. *Internal oblique.* Cut a "window" about one inch square in the external oblique and remove the cut section. Be extremely careful here since the abdominal muscles in the fetus are very thin. The area you have exposed is the internal oblique whose fibers run dorsally and caudally almost at right angles to the direction taken by the fibers of the external oblique. Origin, from the lumbodorsal fascia; insertion, into the linea alba; action, like the external oblique.

12. *Transversus.* With the same care, remove a small portion of the internal oblique. A very thin muscle layer, whose fibers are oriented almost dorsoventrally, will be revealed—this is the transversus. Strip away a small portion of the muscle to expose a shiny membrane, the *peritoneum,* which lines the body cavity or *coelom.* Do not cut the peritoneum since this will permit the escape of a considerable amount of reddish-brown fluid from the body cavity. Origin of the transversus, from the lumbodorsal fascia; insertion, into the linea alba; action, with the obliques.

13. *Rectus abdominis.* Two long strips of muscle lying on either side of the midventral line and covered by the aponeuroses of the obliques and transversus. Slit the aponeuroses to expose the muscle. Origin, from the pubic symphysis; insertion, into the sternum; action, constricts abdomen.

MUSCLES OF THE UPPER FORELIMB
(Figures 3.1, 3.2)

14. *Triceps brachii.* A rather large muscle mass covering almost the entire outer surface of the upper part of the forelimb. It is divisible into three parts, the lateral and long heads, with the medial head lying deeply between the other two. Origin, from the upper end of the humerus; insertion, on the olecranon process of the ulna; action, extends the forelimb.

15. *Biceps brachii.* A rather small spindle-shaped muscle lying along the anterior surface of the humerus. Origin, from the glenoid area; insertion, into the radius; action, flexes the forelimb.

16. *Brachialis.* A small muscle located in the angle between upper and lower arm. Origin, proximal portion of the humerus; insertion, proximal portion of the ulna; action, flexes lower leg.

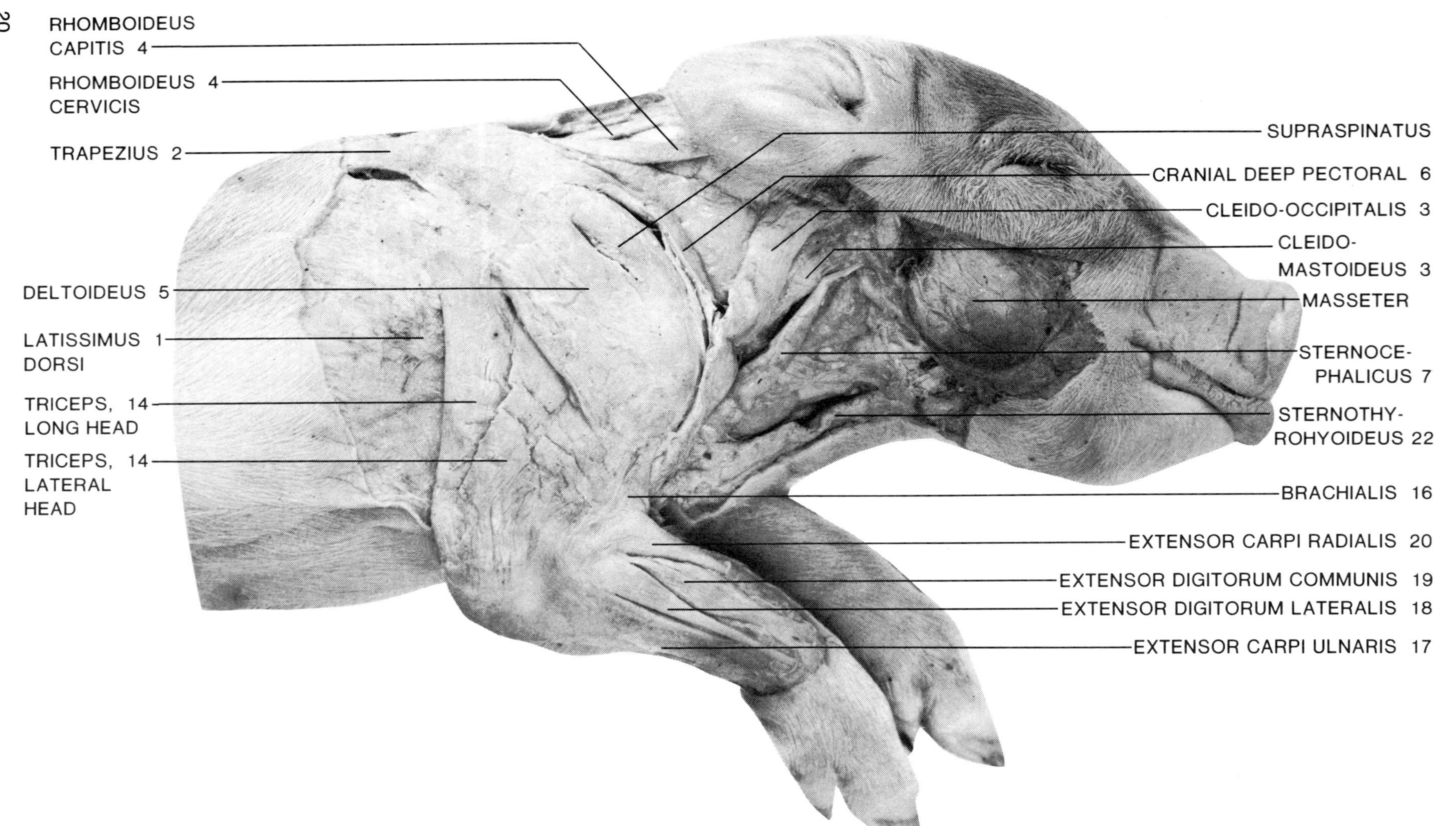

FIGURE 3.2. Superficial muscles of shoulder, neck, and forelimb.

MUSCLES OF THE LOWER FORELIMB
(Figures 3.1, 3.2)

There are a number of muscles of the lower forelimb, most of which are concerned with movements of the foot; not all of these muscles will be identified. Beginning with the caudal edge of the forelimb, identify the following:

17. *Extensor carpi ulnaris.* Locate the *olecranon* process (elbow) to serve as a landmark. Locate a long muscle on the outer caudal surface of the forelimb and slit the fascia between it and the muscle next to it. Origin, lateral portion of distal end of humerus; insertion, by means of a long thin tendon into the fifth metacarpal; action, extends wrist.

18. *Extensor digitorum lateralis.* Similar in appearance to the preceding and lying just cranial to it. Origin, lateral surface of the distal end of the humerus; insertion, by means of a divided tendon into the digits; action, extension of the digits.

19. *Extensor digitorum communis.* A long thin muscle similar to the preceding lying cranial to it along the lateral surface of the arm. Origin, from the lateral surface of the distal end of the humerus; insertion, by means of a divided tendon into the digits; action, in common with the two preceding muscles, extension of the digits.

20. *Extensor carpi radialis.* Lying cranial to the preceding and sometimes partially covered by it. Origin, from the distal end of the humerus; insertion, into the distal end of the radius; action, rotation of the foot.

21. *Brachioradialis.* Cranial to the preceding muscle and often somewhat widely separated from it. Origin, on the humerus; insertion, lower part of the radius; action, rotation of forelimb.

MUSCLES OF THE NECK AND THROAT
(Figure 3.3)

22. *Sternothyrohyoideus.* Two long flat muscles united to each other in the midline of the throat and extending from the sternum to the hyoid bone. Find the junction point of the muscles to each other and slit them apart. Notice the large masses of *thymus* gland lying below the muscles. Origin, from the first costal (rib) cartilage; insertion, into the body of the hyoid; action, moves the hyoid caudally.

23. *Sternothyroideus.* Cut across the bellies of the sternothyrohyoids and expose two small slender muscles. Notice that after taking origin from the sternum, the muscles each separate into two parts and insert at two points, the lateral and ventral surfaces of the *thyroid cartilage* of the *larynx;* action, moves the larynx caudally.

MUSCLES OF THE THIGH
(Figures 3.1, 3.4, 3.5)

24. *Tensor fasciae latae.* The most cranial of the thigh muscles. It is a triangular, thin, short muscle continued ventrally as a sheet of connective tissue, the *fascia latae,* to the patella. Origin, from the ilium; insertion, into the fascia latae; action, tightens the fascia latae and extends the leg.

25. *Gluteus medius.* A thick muscle mass almost completely covered by the tensor fasciae latae and the fascia of the hip. Split the fascia and remove it to expose the underlying gluteus medius. Origin, lumbodorsal fascia and fascia of the hip; insertion, into the proximal end of the femur; action, abducts thigh.

26. *Biceps femoris.* A large muscle comprising most of the caudal half of the lateral surface of the thigh. Origin, from the ischium; insertion, lower end of the femur and upper part of the tibia; action, abducts thigh and flexes shank.

27. *Vastus lateralis.* This muscle comprises most of the cranial part of the thigh but the fascia of the tensor fasciae latae must be removed before the muscle will be seen. Origin, head of the femur; insertion, into the patella; action, extends the shank.

28. *Vastus medialis.* On the cranial medial portion of the thigh. Origin, from the head of the femur; insertion and action, with the vastus lateralis and rectus femoris.

29. *Rectus femoris.* Along the cranial edge of the thigh you will find a junction between the vastus lateralis and the vastus medialis on the medial side of the thigh. Separate the two vasti muscles to expose a long spindle-shaped muscle, the rectus femoris. Origin, from the ilium; insertion, with the vastus lateralis; action, extends the shank.

30. *Vastus intermedius.* Separate the vastus lateralis from the rectus femoris and look deeply between them. A small muscle will be seen lying on the shaft of the femur. Origin, shaft of the femur; insertion, into the patella; action, extends the shank. The three vasti muscles, together with the rectus femoris, are known as the *quadriceps femoris.* The four muscles are *synergists* (working together).

31. *Semitendinosus.* Cut across the belly of the biceps femoris and reflect the two halves. The semitendinosus will be seen as a thick band of muscle under the caudal edge of the biceps. Origin, first caudal vertebrae and ischium; insertion, into the upper end of the tibia; action, flexes the shank.

32. *Semimembranosus.* This is the most caudally placed muscle of the thigh and lying close to the semitendinosus. Origin, ischium; insertion, distal

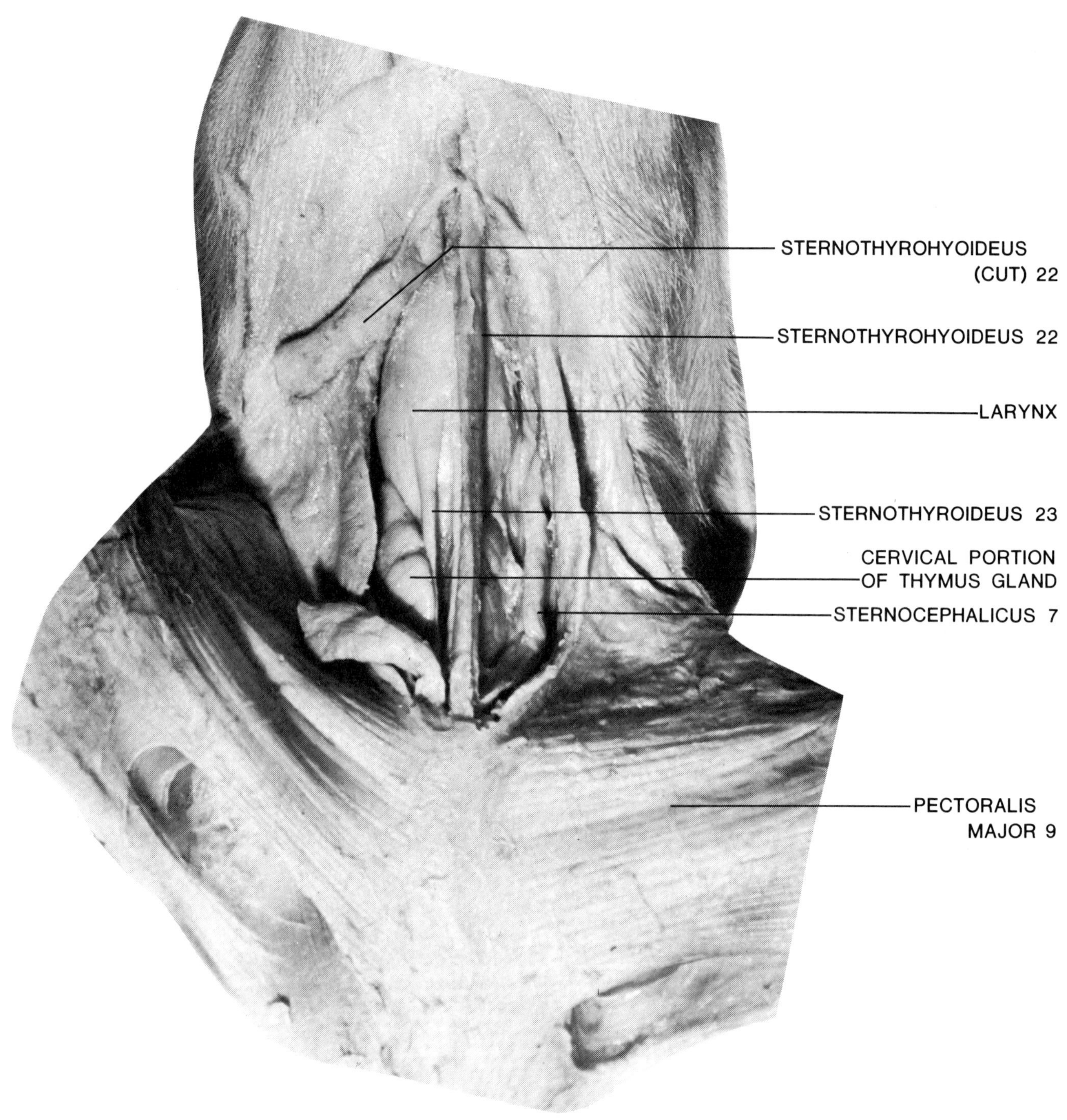

FIGURE 3.3. Muscles of the throat and upper thorax.

end of the femur and proximal end of the tibia; action, extends the thigh.

33. *Sartorius.* A thin triangular muscle on the medial surface of the thigh. Origin, two heads from the ilium with the external iliac artery and vein lying between the two heads; insertion, into the proximal end of the tibia; action, adducts the femur.
34. *Gracilis.* A broad thin muscle on the caudal half of the medial surface of the thigh. Origin, from the pubis and ischium; insertion, proximal end of the tibia; action, adducts the thigh.
35. *Adductor.* On the medial surface of the thigh and lying between the sartorius and gracilis. Origin, from the ischium and pubis; insertion, distal end of femur; action, adducts the hindlimb.

MUSCLES OF THE SHANK
(Figures 3.1, 3.4)

Remove the tough fascia covering the shank muscles and identify the following:

36. *Gastrocnemius.* The large muscle on the caudal surface of the shank consists of a lateral head and a medial head. Origin, distal end of the femur; insertion, by the Tendon of Achilles to the calcaneus; action, extends the foot.
37. *Soleus.* Lying under the lateral head of the gastrocnemius and usually closely attached to it. Origin, proximal end of the fibula; insertion and action, with the gastrocnemius.
38. *Digital flexors.* Next cranial to the gastrocnemius. Origin, proximal ends of the tibia and fibula; insertion, onto the digits; action, flexes digits.
39. *Digital extensors.* Origin, proximal ends of the tibia and fibula; insertion, metatarsals; action, extends the digits.
40. *Peroneus longus.* Origin, proximal ends of the tibia and fibula; insertion, metatarsals; action, flexes ankle.

41. *Peroneus tertius.* Origin, lateral surface of the distal end of the femur; insertion, third metatarsal; action, flexes foot.

MICROSCOPIC STRUCTURE OF MUSCLES
(Figures 3.6, 3.7, 3.8)

The three types of muscles of the mammalian body are cardiac, smooth, and skeletal or striated. Cardiac muscle is found exclusively in the heart; smooth muscle in the hollow organs of the body such as stomach, large and small intestine, uterus, urinary bladder, blood vessels, and gall bladder; striated muscle occurs in bundles attached to bone by means of connective tissue and is responsible for the gross movements of the body.

An examination of microscope slides of the three muscle types reveals some of their detailed structure. This is only a brief survey and no attempt is made to include all of the details of microscopic anatomy.

1. Cardiac Muscle (uninucleated) (fig. 3.6). These are muscle fibers characterized by a branching arrangement and by the presence of darker bands, the *intercalated discs,* lying at right angles to the muscle fibers. These discs are the ends of the muscle cells.
2. Smooth Muscle (uninucleated) (fig. 3.7). These muscle cells are typically spindle-shaped and occur in flat sheets or layers. Each cell has a prominent nucleus and there are no striations, which are characteristic of skeletal muscle.
3. Striated (skeletal) Muscle (multinucleated) (fig. 3.8). These muscles comprise the bulk of the musculature of the body and are attached to the skeletal system to provide for locomotion and movement. The characteristic feature of these muscles are the cross striations, which are composed of the dark A Bands alternating with the light I Bands. Each muscle fiber is surrounded by a thin layer of connective tissue called *endomysium.*

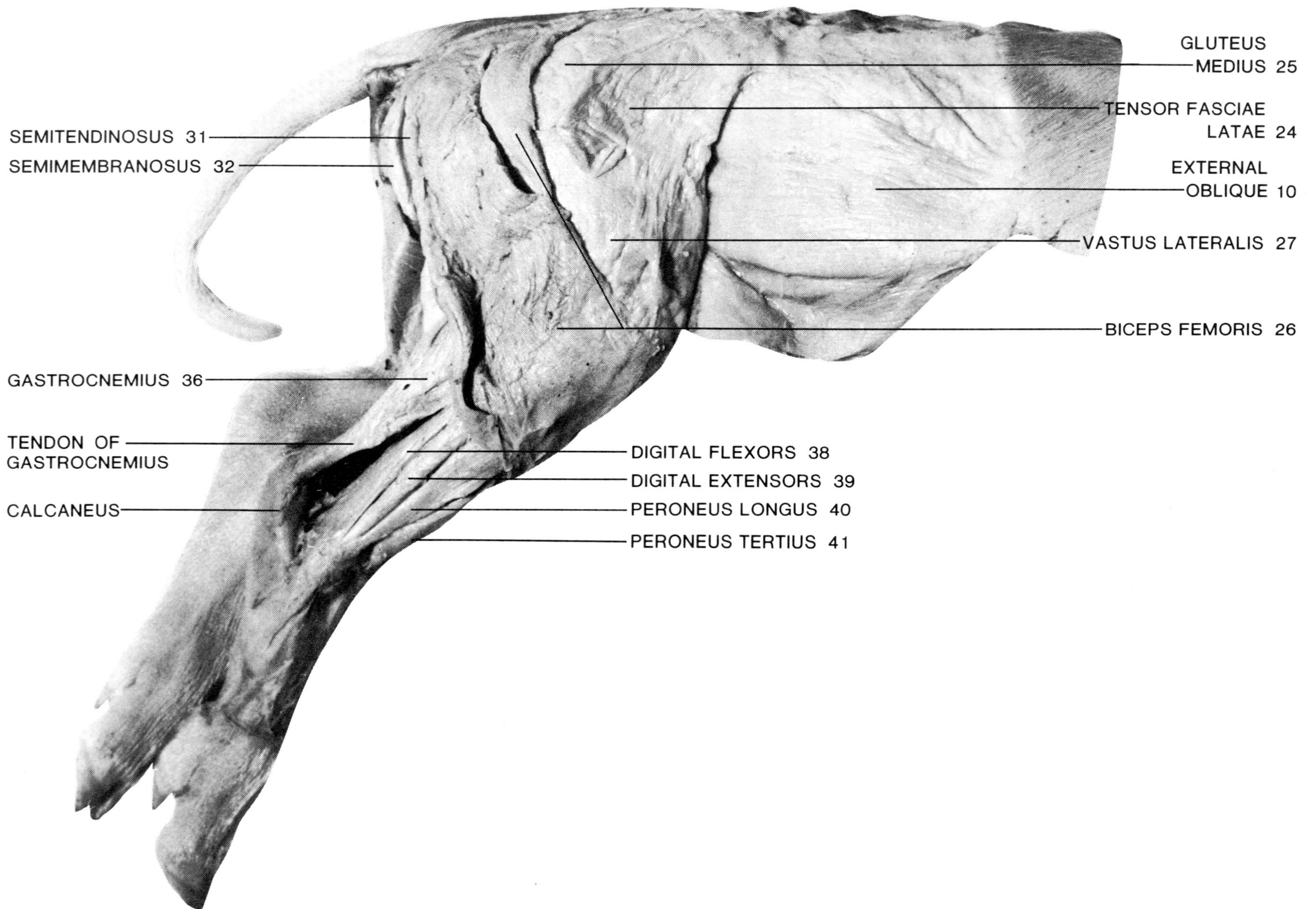

FIGURE 3.4. Superficial muscles of thigh and leg, lateral view.

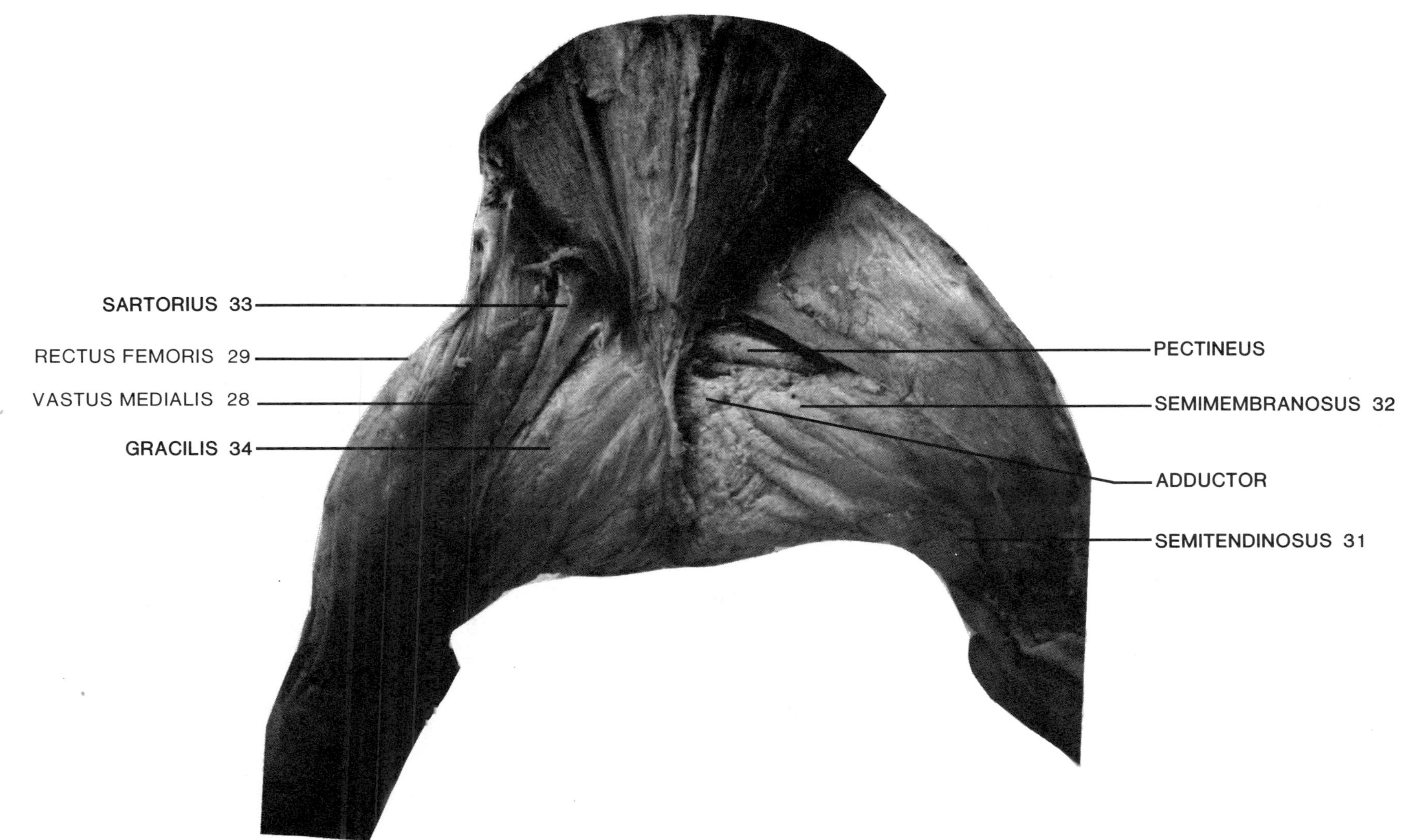

FIGURE 3.5. Muscles of medial surface of the thigh, ventral view.

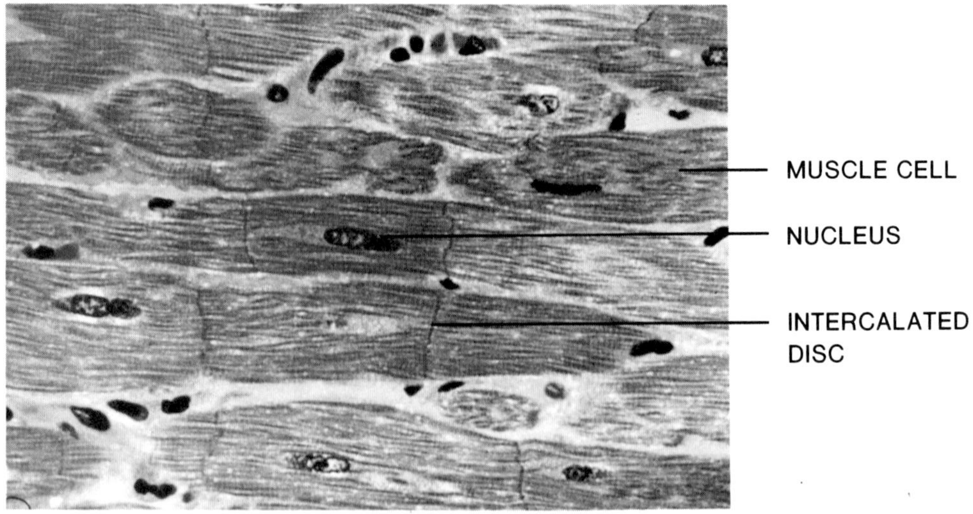

FIGURE 3.6. Longitudinal section of cardiac muscle (photomicrograph ×160).

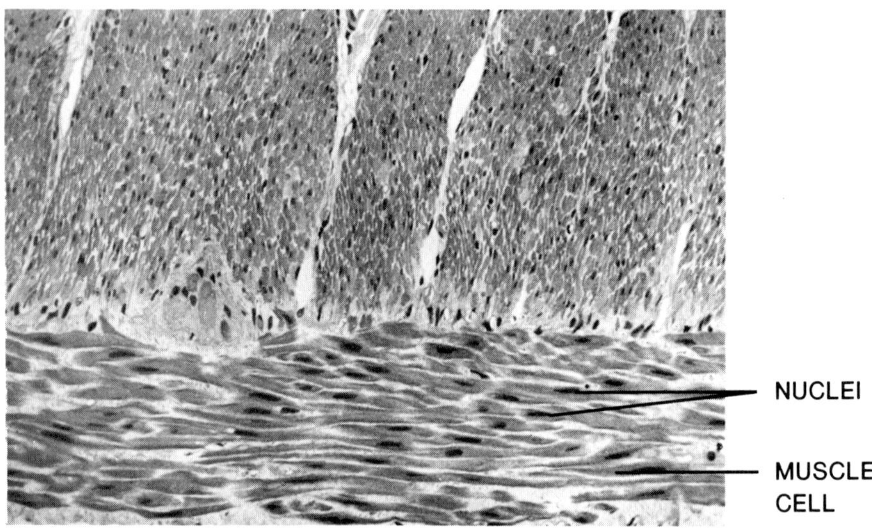

FIGURE 3.7. Longitudinal section of smooth muscle (photomicrograph ×400).

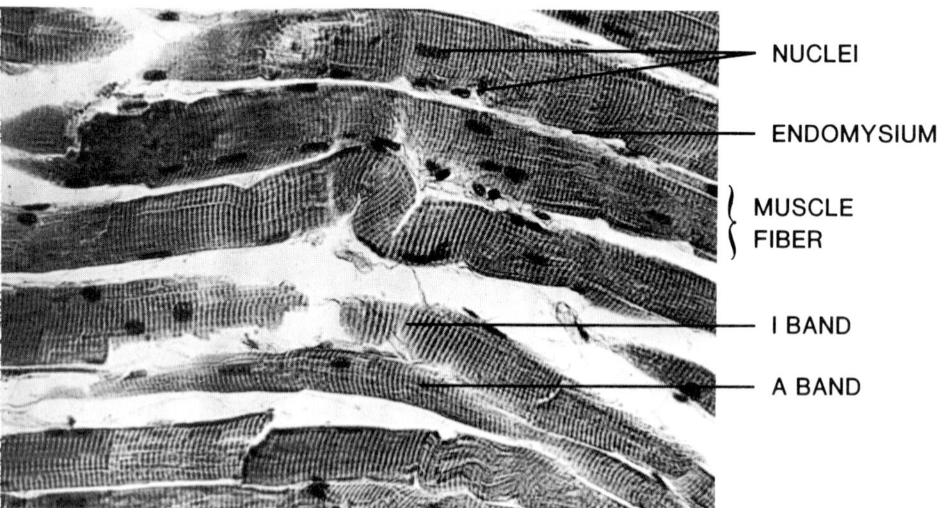

FIGURE 3.8. Longitudinal section of striated muscle (photomicrograph ×160).

Chapter 4
General Internal Anatomy

(Figures 4.1a, 4.1b, 4.2, 4.3)

CHAPTER OBJECTIVES

1. To expose and identify the major coelomic or body cavities.
2. To identify the organs associated with the thoracic and abdominal cavities.

INTRODUCTION

The two major body (coelomic) cavities are the thoracic (pleural) and abdominal (peritoneal). A thin tissue sheet lines the internal surface of these cavities and also encloses some of the organs contained within. Separating the two cavities is a musculo-tendinous structure, the diaphragm.

The thoracic cavity is lined with *parietal pleura.* The organs lying within this space are the lungs, which are enclosed in *visceral pleura,* the heart, blood vessels such as the aorta and venae cavae, the trachea, and the esophagus.

The abdominal cavity is lined with a shiny membrane, the *peritoneum,* and contains such viscera as the stomach, large and small intestine, liver, gall bladder, pancreas, uterus, ovaries, spleen, and major blood vessels and nerves. The kidneys appear to lie within the cavity but, in a strict sense, they lie outside the peritoneal lining. Their position is known as *retro-peritoneal.* The peritoneum is continuous with a well-developed tissue sheet, the *mesentery,* which suspends organs such as the stomach and intestine from the dorsal body wall.

Posteriorly, the abdominal cavity is continuous with a smaller pelvic cavity containing the internal portions of the urinary and reproductive systems and the continuation of the digestive system, the rectum.

DIRECTIONS FOR DISSECTION
(Figure 4.1)

All of the viscera, except when otherwise specifically directed, are to be left intact in the body. For the following work, use the outline drawing of the pig, which shows where the incisions are to be made.

Place the pig, dorsal side down, in the dissecting pan. Tie a cord around the lower end of the forelimbs and hindlimbs on one side of the body, bring the cords under the pan to the opposite side, and tie them to the forelimbs and hindlimbs on that side.

If your specimen is a *female,* deepen the incision you have made previously in studying the muscles. Begin at the small hairy papilla on the upper part of the throat (1) and continue posteriorly to the umbilical cord; cut around both sides of the cord (2), then direct the two cuts into one just caudal to the cord (3 ♀) and continue back between the hindlegs (fig. 4.1a).

If your specimen is a *male,* proceed as with the female down to the umbilical cord. Instead of bringing the cuts together behind the cord, continue the *two* incisions caudally between the hindlegs (3 ♂). The reason for doing so is to avoid cutting the penis which lies under the skin in the midventral line (fig. 4.1a).

For *both sexes,* continue to deepen the incisions until the body cavity is reached. If the cavity is filled with a dark fluid, flush it out with water. Just anterior to the hindlegs (4) and just caudal to the front legs (5) *cut into and through the body wall* in order to make lateral flaps that can be pinned down out of the way. Free the *diaphragm* from the body wall by cutting the edge of the diaphragm where it is in contact with the body wall (6).

You will probably have noticed by this time that the umbilical cord and the flesh immediately around it cannot be laid back freely on the body. Look for a dark tubular structure extending from the umbilical cord anteriorly into the *liver;* this is the *umbilical vein.* Tie a small piece of cord around the vein in two places and then cut through the vein between the cords. The umbilical cord with its attached strip of flesh can now be pulled back. The cords tied around the umbilical vein will serve to identify the vein later on.

THORACIC VISCERA
(Figures 4.2, 4.3)

In this, as well as the following section, make only general observations with respect to size, color, and relative position of the organs.

1. *Diaphragm.* A musculo-tendinous sheet separating the *pleural* cavity containing the heart and lungs from the larger caudal *peritoneal* cavity containing the visceral organs such as the stomach, liver, small and large intestine, spleen, urinary bladder, and internal reproductive organs.

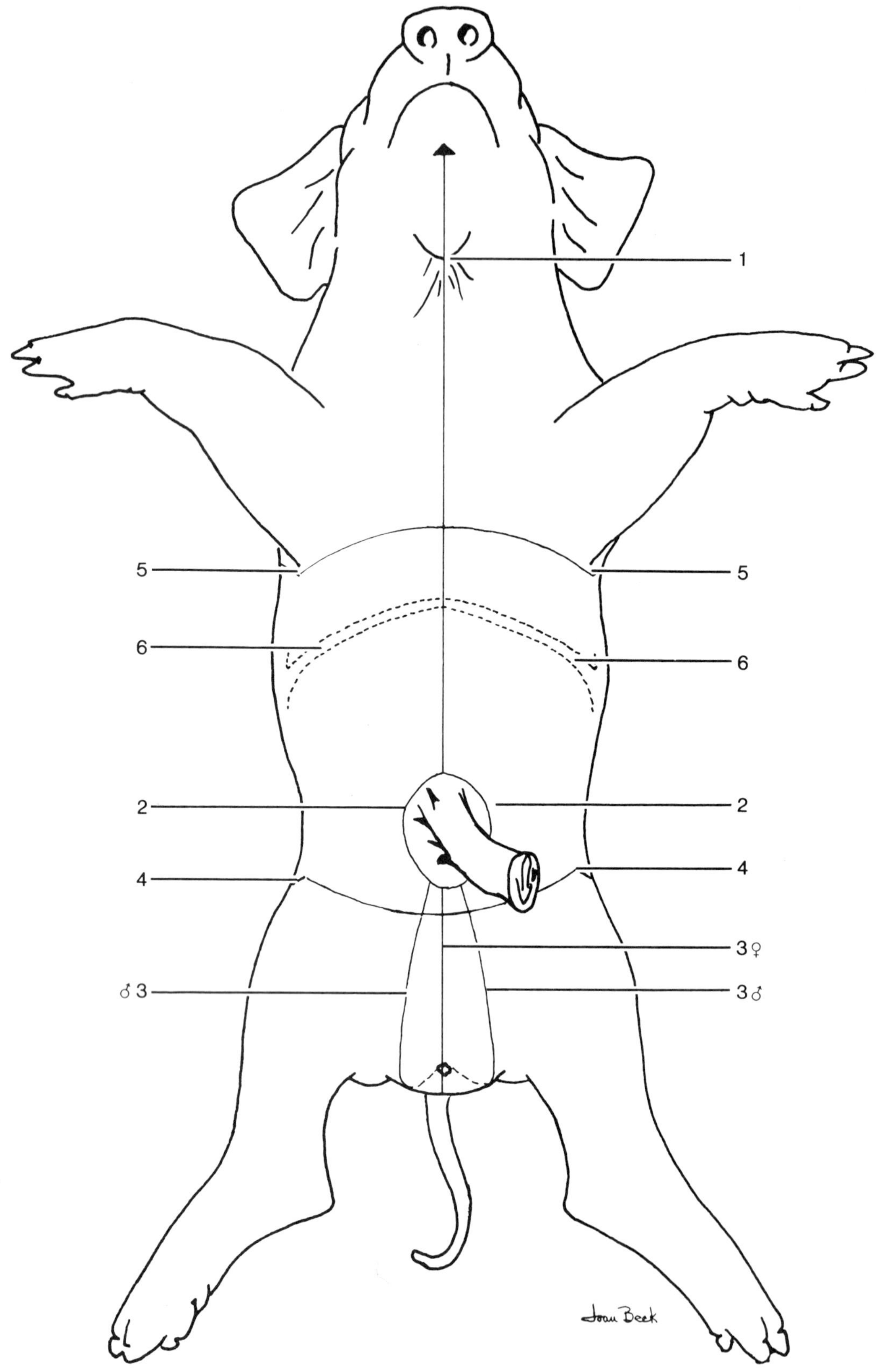

FIGURE 4.1a.   Incisions to be made, in numerical order.

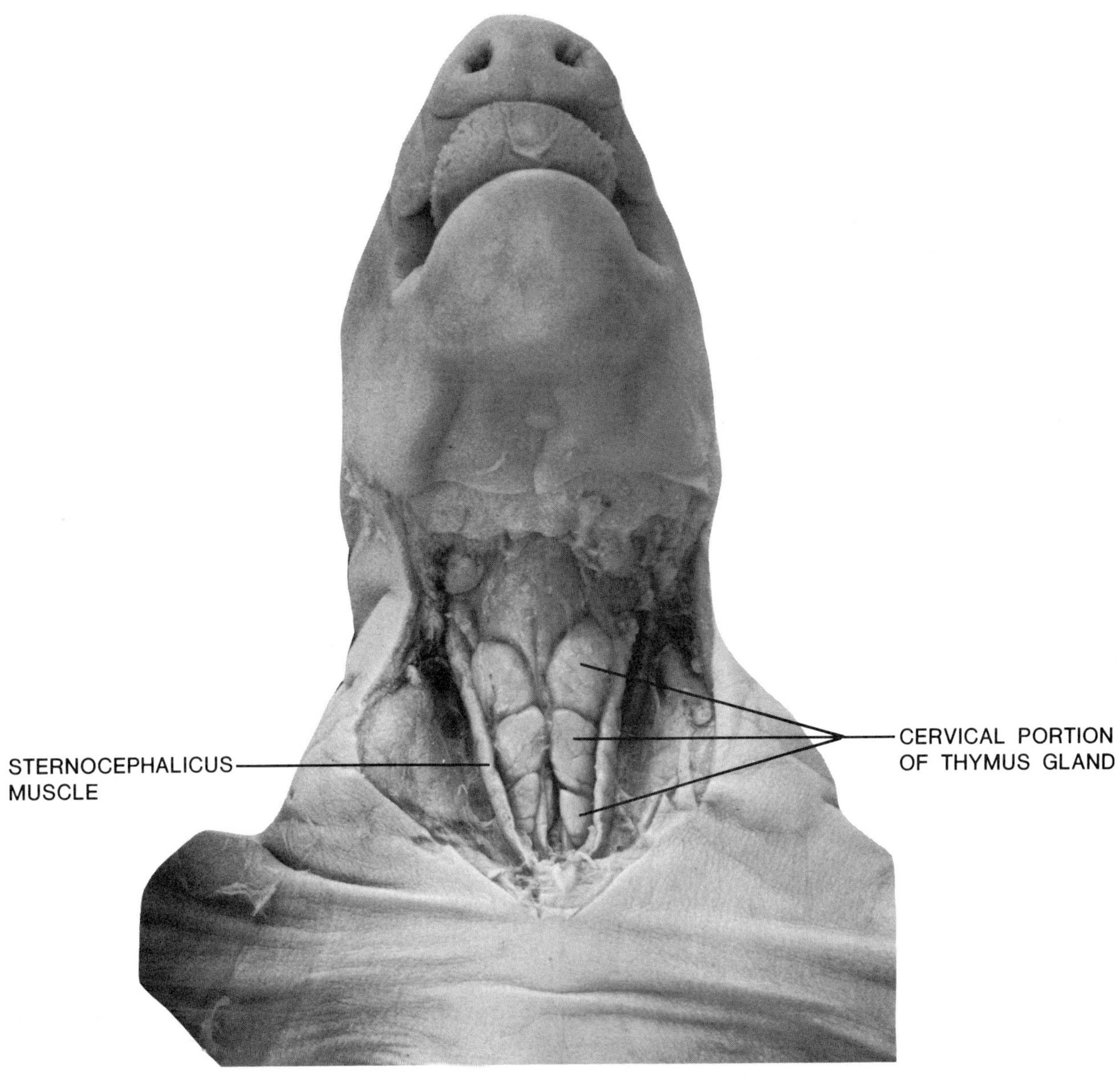

FIGURE 4.1b. Cervical portion of thymus gland, ventral view.

2. *Thymus glands*. Large elongated masses extending from the throat region down over the cranial portion of the heart (figs. 4.1b, 4.2).
3. *Thyroid gland*. An oval reddish gland near the base of the throat.
4. *Heart and lungs*. Before you proceed further with the opening of the pleural cavity, notice the thin membrane extending from the ventral wall of the chest down to and around the heart; this is the *mediastinal septum*. Note also the *pericardial membrane* immediately around the heart.

THYMUS AND THYROID GLANDS

Cut through the skin of the throat in the midventral line, then cut laterally just anterior to the larynx and to the sternum, free the skin flaps from underlying tissue and remove them. Two long muscle bands, the *sternothyrohyoideus,* will be seen. Remove the muscles to reveal a large lobed structure on either side of the midventral line. These are the cervical portions of the thymus gland consisting of three or four lobes on either side. Now extend the midventral incision in the throat caudally to the midventral thoracic region. Here it will be necessary to carefully cut through the sternum removing it piece by piece until a glandular mass is exposed lying ventral to the pericardium of the heart. This is the thoracic portion of the thymus gland.

To expose the thyroid gland (see figure 4.3), spread apart the lobes of the cervical portion of the thymus. This will reveal a bi-lobed reddish-brown thyroid gland with one lobe on each side of the trachea.

ABDOMINAL VISCERA
(Figures 4.2, 4.3)

The organs named below are listed only for the purpose of general observation. Detailed descriptions will appear in following chapters.

1. *Liver*. The large, brownish lobed structure lying in the cranial end of the abdominal cavity just caudal to the diaphragm.
2. *Stomach*. A large sac situated in the left side of the abdominal cavity and almost entirely covered by the liver.
3. *Spleen*. An elongated reddish-brown organ attached by means of the *greater omentum* to the *greater curvature* of the stomach.
4. *Small intestine*. Filling most of the abdominal space caudal to the stomach and liver.
5. *Large intestine* or *spiral colon*. A compact darker mass of coils lying in the left part of the body cavity and partially overlapped by the spleen.
6. *Cecum*. Lift the coils of the small intestine forward to reveal a small sac or pouch at the junction of the small and large intestines.
7. *Urinary (Allantoic) bladder*. The large sac situated between the two umbilical arteries at the caudal end of the body.
8. *Peritoneum*. The smooth shiny membrane lining the abdominal cavity. This is the *parietal peritoneum* surrounding the *coelom* or body cavity. Note that the parietal peritoneum extends from the body wall in the dorsal region and surrounds the internal organs as a suspensory sheet; this sheet is a *mesentery* and its continuation around the internal organs is known as the *visceral peritoneum*.

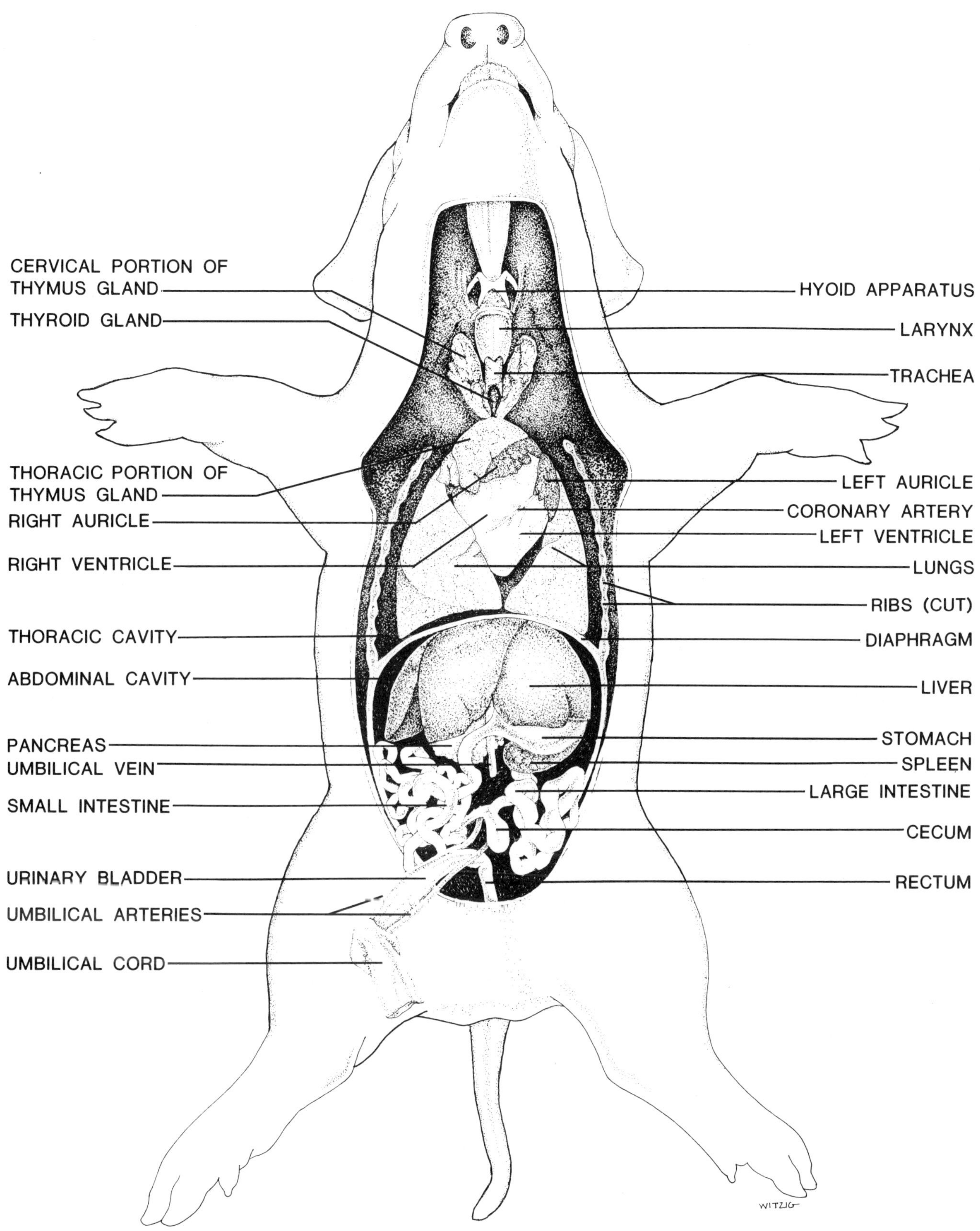

FIGURE 4.2. **General internal anatomy, ventral view.**

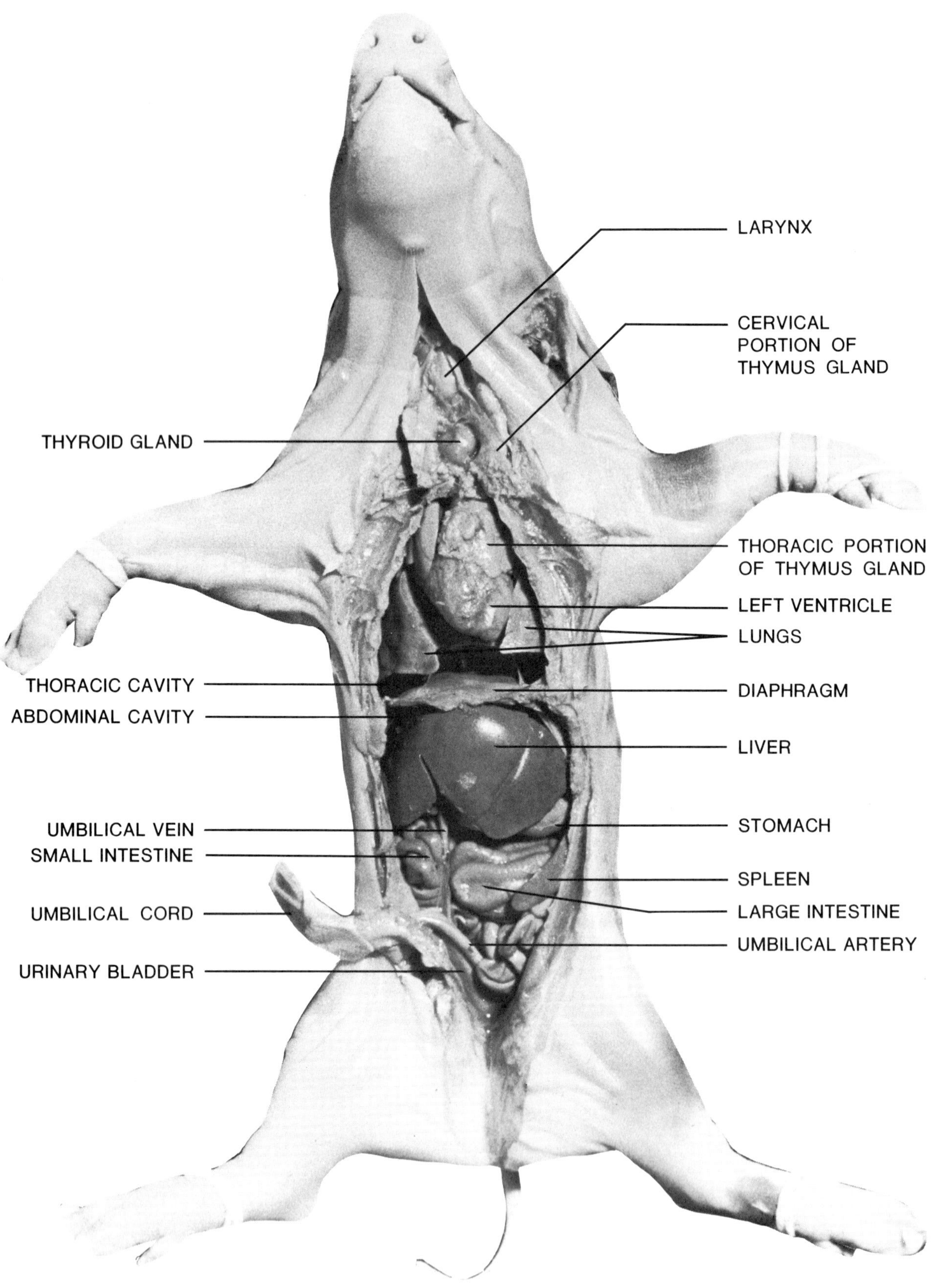

FIGURE 4.3. General internal anatomy, ventral view.

32

Chapter 5
Digestive System

(Figures 5.1, 5.2, 5.3, 5.4a, 5.4b, 5.5a, 5.5b, 5.6, 5.7)

CHAPTER OBJECTIVES

1. To identify the gross features of the digestive system: mouth, pharynx, esophagus, stomach, small intestine, large intestine, and associated glands such as salivary, liver, and pancreas.
2. To identify the microscopic structure of the stomach and small intestine.

INTRODUCTION

The function of the digestive system is to reduce raw food materials to a condition where such materials can be absorbed into and utilized by the organism. This is accomplished mechanically by the teeth and chemically by enzymes produced by salivary glands and glands of the stomach, small intestine, and pancreas.

SALIVARY GLANDS
(Figures 5.1, 5.2)

There are three pairs of salivary glands, the *parotid, mandibular (submaxillary),* and *sublingual.* Remove the skin and muscle layer from one side of the face and neck to expose a large, rather dark triangular gland, the parotid. The differentiation between muscle tissue and glandular tissue is the presence of fibers in muscle, whereas the parotid gland has the appearance of being composed of many small nodules. The parotid lies just ventral to the base of the ear and extends to the caudal border of the masseter muscle. The duct that drains this gland is the *parotid (Stensen's) duct,* arising from the upper part of the gland on its ventral surface. It proceeds ventrally around the lower edge of the masseter and opens into the mouth cavity near the upper fourth premolar tooth.

The mandibular gland is large and somewhat lobed. It is located just caudal to the masseter muscle and partially covered by the cranial portion of the parotid. The mandibular duct emerges from the cranial edge of the gland, proceeds anteriorly, and opens on the floor of the mouth.

The sublingual gland is long and rather slender with only its caudal part readily visible surrounding the proximal portion of the mandibular duct. The sublingual duct also opens on the floor of the mouth.

The parotid gland secretes a watery fluid, the mandibular and sublingual glands a more viscous fluid. These secretions function in maintaining the mucous membranes of the mouth and pharynx in a moist condition and in lubricating food for ease in swallowing. In addition, the glands produce enzymes which initiate carbohydrate digestion by reducing starch to maltose and glucose.

MOUTH CAVITY
(Figure 5.3)

With the bone shears, cut back through the corners of the mouth and continue well posterior to the masseter muscle. If necessary, continue the cut until the lower jaw can be dropped and the interior of the mouth examined. Observe the space between the lips and the teeth (or gum if teeth are not yet present); this is the *vestibule.* The larger space of the mouth cavity is the *oral cavity.*

1. *Tongue.* An elongated muscular sac attached to the floor of the oral cavity by a membrane, the *frenulum linguae.* On the surface of the tongue are a number of *papillae* containing *taste buds.* The majority of these papillae are called *fungiform* and are situated along the edges of the tongue in the caudal region and around the tip of the tongue. Three other types of papillae are present but they are small and few in number and will not be identified.
2. *Teeth.* As a group, mammals possess the following types of teeth: *incisors* for cutting, *canines* for tearing, and *premolars* and *molars* for chewing and grinding. The full set of temporary or *deciduous* teeth may or may not be present in your specimen depending upon its age. If your specimen has no teeth in evidence, cut into the gum and observe whether developing teeth are present.
3. *Hard palate.* Composed of bone and covered with a mucous membrane which is ridged in a number of transverse folds.
4. *Soft palate.* The caudal continuation of the mucous membrane covering the hard palate.

PHARYNX

1. *Isthmus of the Fauces.* The opening of the oral cavity into the pharynx.
2. *Nasopharynx.* Slit the soft palate through the midline and remove both halves. The space revealed is the nasopharynx.
3. *Esophageal aperture.* Find the esophagus, a soft muscular tube lying on the dorsal surface of the trachea. Slit the esophagus, insert a probe into it and run the probe back up into the mouth and note the place where it emerges.
4. *Glottis.* The opening into the *trachea* identified by the presence of a small white tab of cartilage, the *epiglottis,* at the point of entrance.

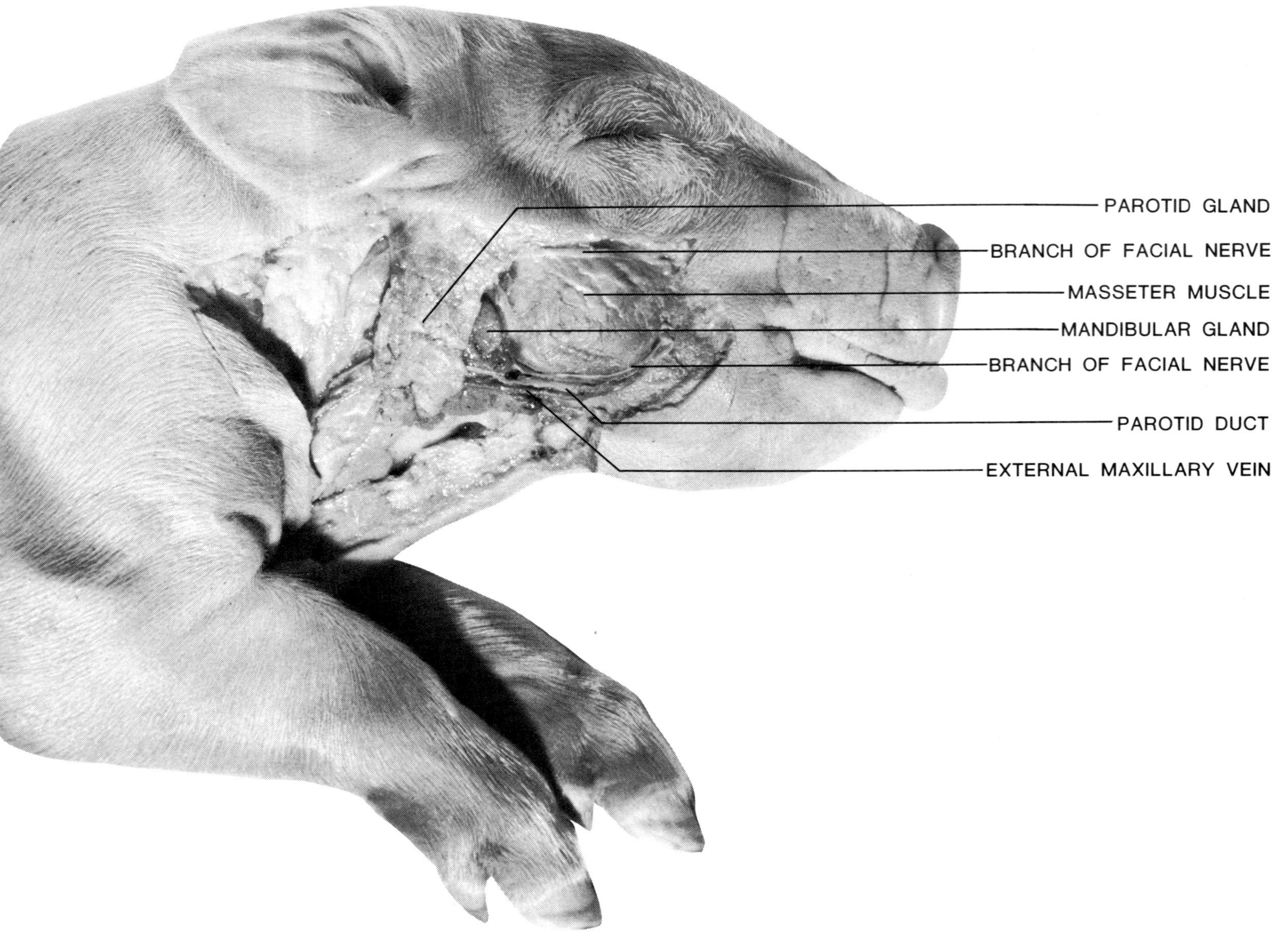

FIGURE 5.1. Salivary glands (sublingual not shown), lateral view.

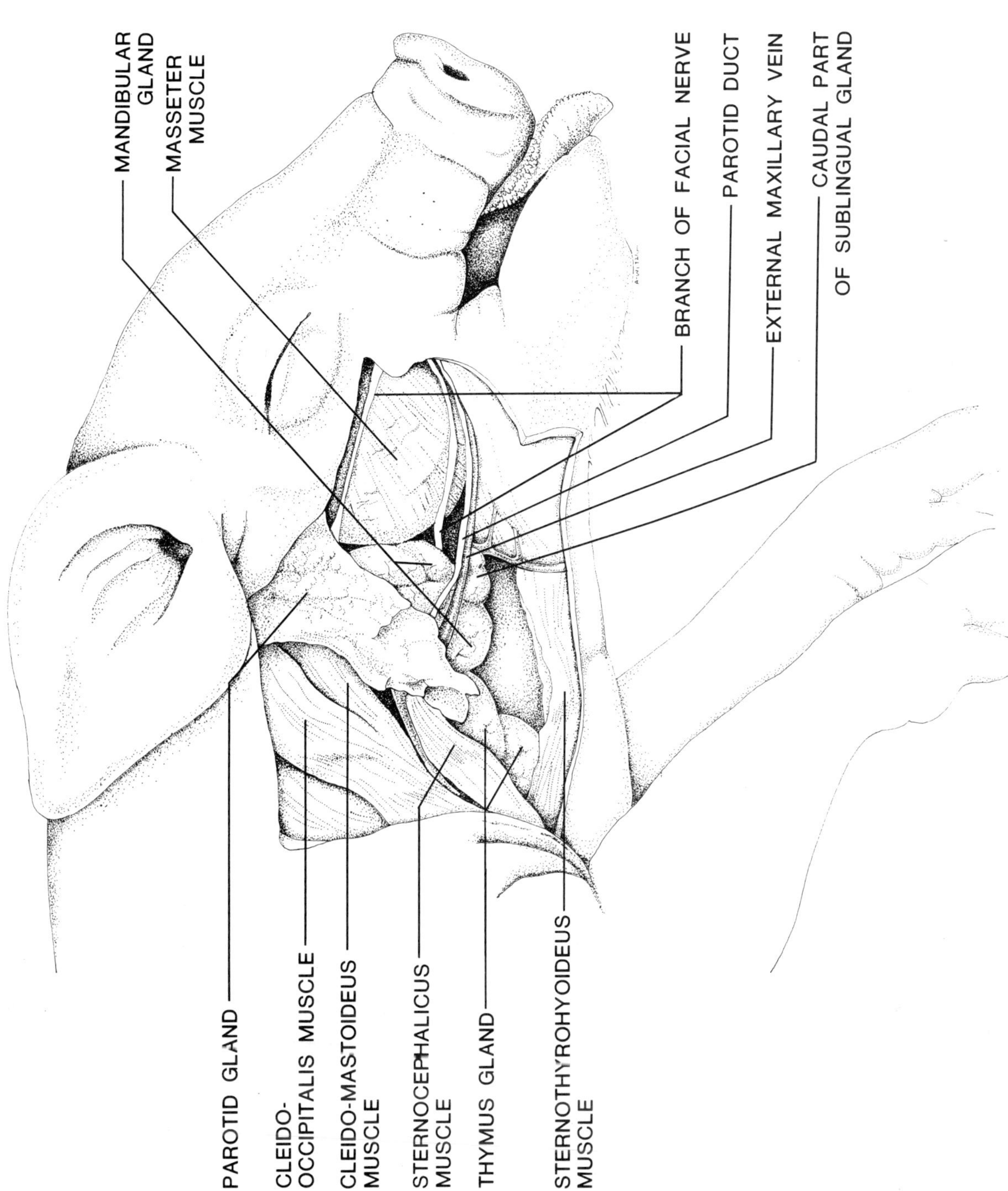

FIGURE 5.2. Salivary glands, lateral view.

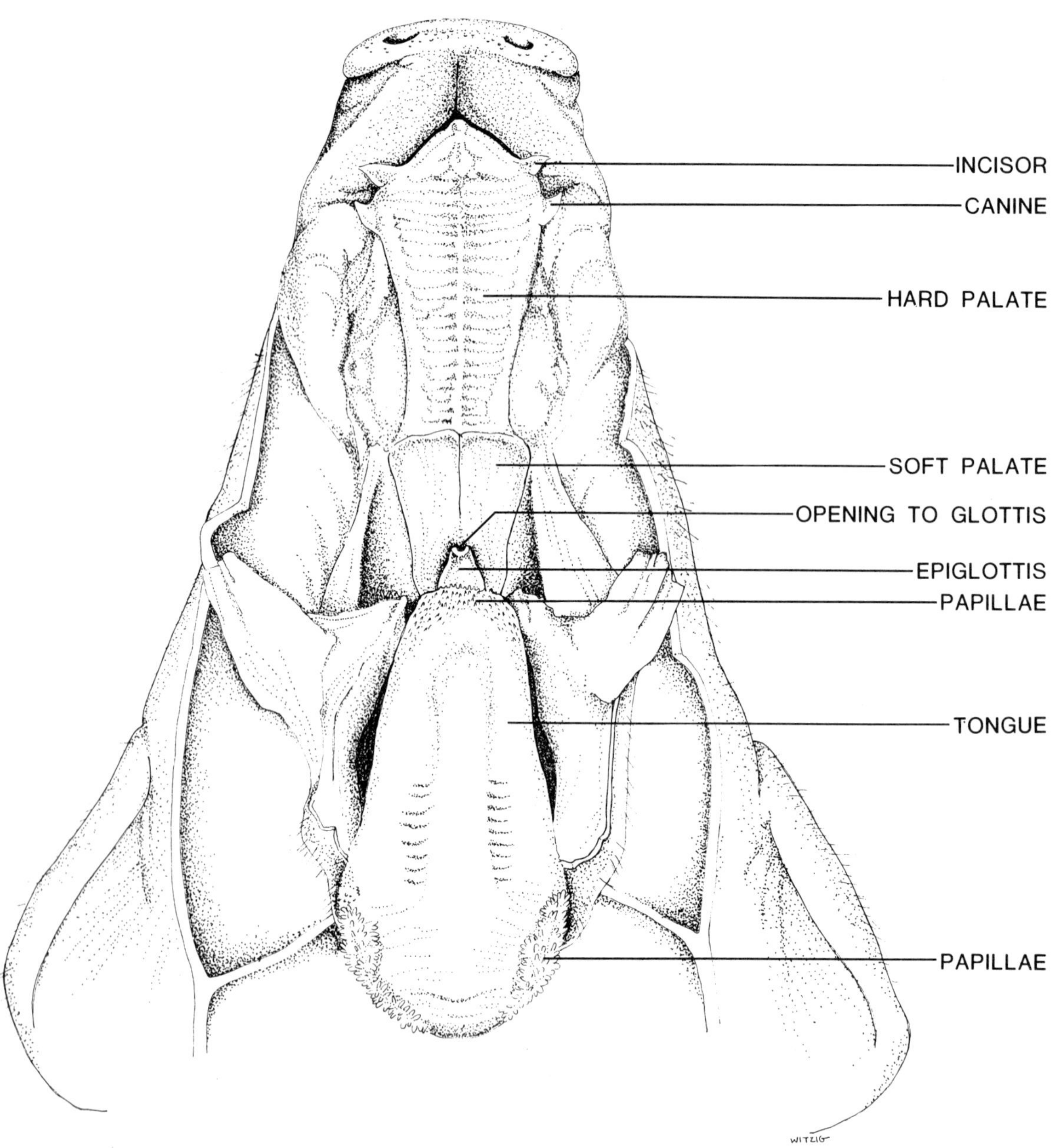

FIGURE 5.3.   Interior of mouth.

1. *Esophagus.* The muscular tube from pharynx to stomach. Notice its close association with the dorsal surface of the trachea. Turn to the abdominal cavity and observe the penetration of the diaphragm by the esophagus before the latter enters the cranial end of the stomach.

2. *Stomach.* Remove the stomach from the body cavity by cutting through the duodenum and esophagus. On the external surface of the stomach identify the *greater curvature,* the longest convex border; the *lesser curvature* opposite the greater curvature; the *diverticulum,* a triangular, flattened protuberance near one end of the stomach; the *esophagus* entering the stomach on the lesser curvature near the diverticulum; the *duodenum* arising from the stomach near the entrance of the esophagus. Cut across the duodenum close to the stomach and note the strong muscle ring which is exposed. This is the *phyloric sphincter* which serves to close off the lumen of the stomach from the duodenum except when food is passing through.

   Cut a frontal section through the stomach, including the stubs of the esophagus and duodenum. On the internal surface a number of regions may be defined, although not all of them are as well developed as they will be in the adult pig. The space within the diverticulum is called the *diverticulum ventriculi.* Next to it is the region into which the esophagus enters—this is the *pars proventricularis.* Next is a region known as the *pars glandularis* whose boundaries are not well defined. The fourth region is the *pars pylorica* from which the duodenum emerges.

3. *Duodenum.* The first portion of the small intestine leading away from the pyloric end of the stomach. The pancreas lies between the stomach and duodenum.

4. *Jejunum and Ileum.* These two regions of the small intestine comprise the rest of the alimentary tract as far as the large intestine and cannot be differentiated from one another. Cut out a small segment of the intestine, about an inch in length, open it longitudinally, flatten it out, and examine it *under water.* Observe the velvety appearance of the internal lining composed of numerous fingerlike projections, the *villi* (see fig. 5.6).

5. *Colon.* This is the first part of the large intestine and is seen in the fetal pig as a compact coiled mass. The ileum opens into the colon at an oblique angle and in such a manner as to form a blind pouch, the *cecum,* at the beginning of the colon;
it can be seen by pushing the coiled mass to one side. Cut into the cecum at about the point where the ileum enters; wash out the contents and look for the *ileocecal* valve.

6. *Rectum.* The caudal part of the large intestine, continuing from the colon as a dark straight tube, and terminating in the anal opening. If it has not already been done, cut through the cartilage of the pelvic girdle and free the rectum from the structures associated with it, i.e., the *vagina* in the female and the *urethra* in the male.

7. *Liver.* The large brownish gland at the cranial end of the peritoneal cavity. It is divided into four main lobes, the right lateral, right medial, left lateral, and left medial. Press the intestinal coils to the right and identify the small caudate lobe of the liver attached to the caudal face of the right lateral lobe.

8. *Gall bladder.* Not a gland but a storage sac for *bile* secreted by the liver. In fresh specimens it has a greenish appearance, but in pigs that have been preserved in formalin, the sac is shrunken and almost colorless. The *cystic duct* from the gall bladder and the hepatic duct from the liver unite and empty into the duodenum as a *common bile duct* (fig. 7.5). Trace the common bile duct from the liver to the duodenum, removing the fatty material that covers it but being careful not to injure the *hepatic portal vein* that runs in company with the bile duct. Note that the bile duct empties into the duodenum through a small papilla, the *duodenal papilla.*

9. *Pancreas.* An elongated granular mass lying in the angle between the curve of the stomach and the duodenum. Do not attempt to find the *pancreatic duct* since it is too small to be dissected satisfactorily.

MICROSCOPIC ANATOMY OF THE SMALL INTESTINE
(Figures 5.6, 5.7)

Study a slide showing a microscopic cross section of the small intestine. Identify the following layers: *mucosa* composed principally of the *villi,* small fingerlike projections into the *lumen* or cavity of the intestine; each villus shows numerous mucous-secreting *goblet cells.* Underlying the mucosa is the *muscularis mucosae,* consisting of smooth muscle fibers and a network of elastic fibers. Next is a layer called the *submucosa,* composed of dense connective tissue. This is followed by a wide band of *circular muscles,* then a layer of *longitudinal muscles.* The outermost covering of the small intestine is tough connective tissue, the *serosa.*

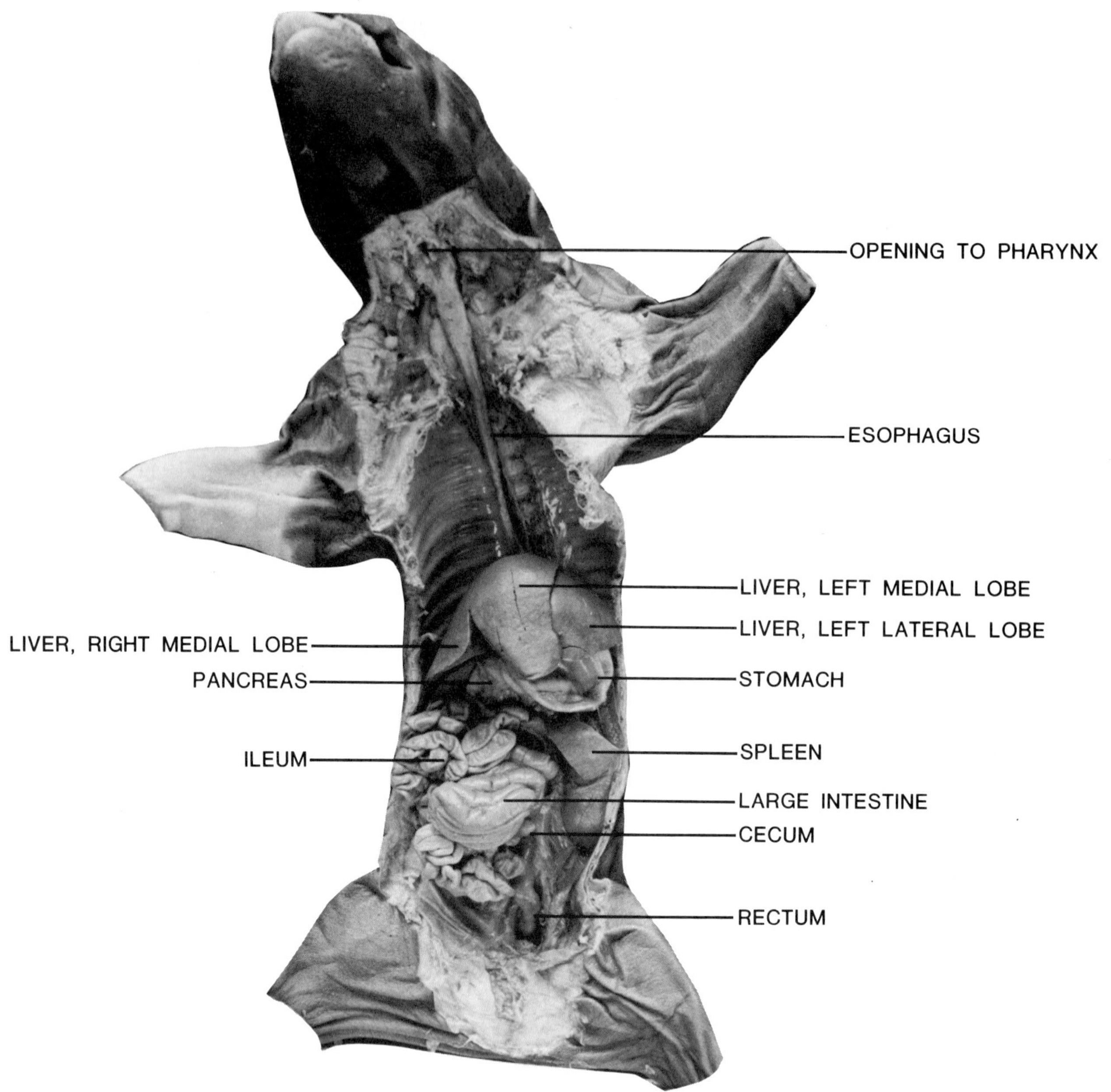

FIGURE 5.4a. Digestive system, ventral view.

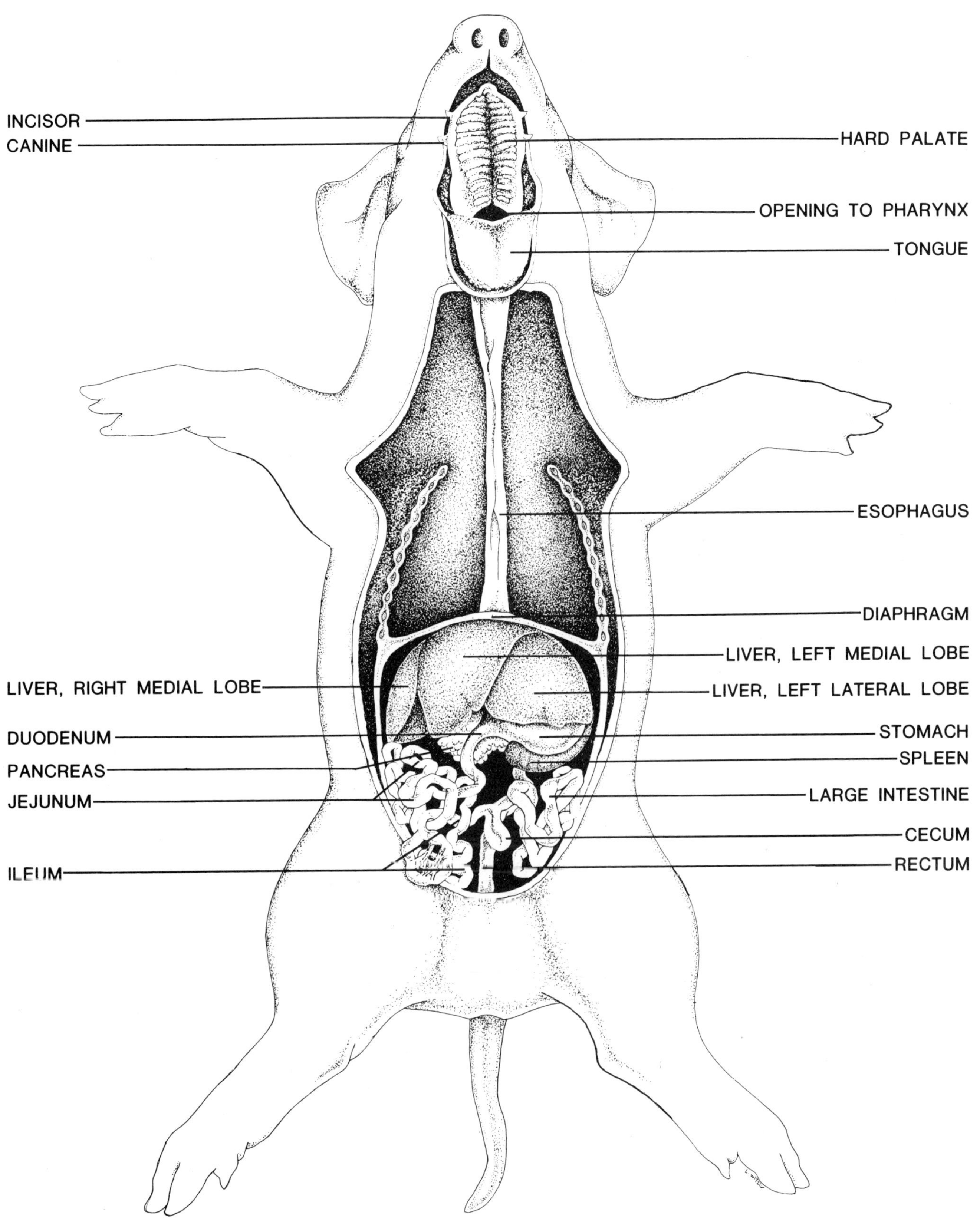

FIGURE 5.4b. Digestive system, ventral view.

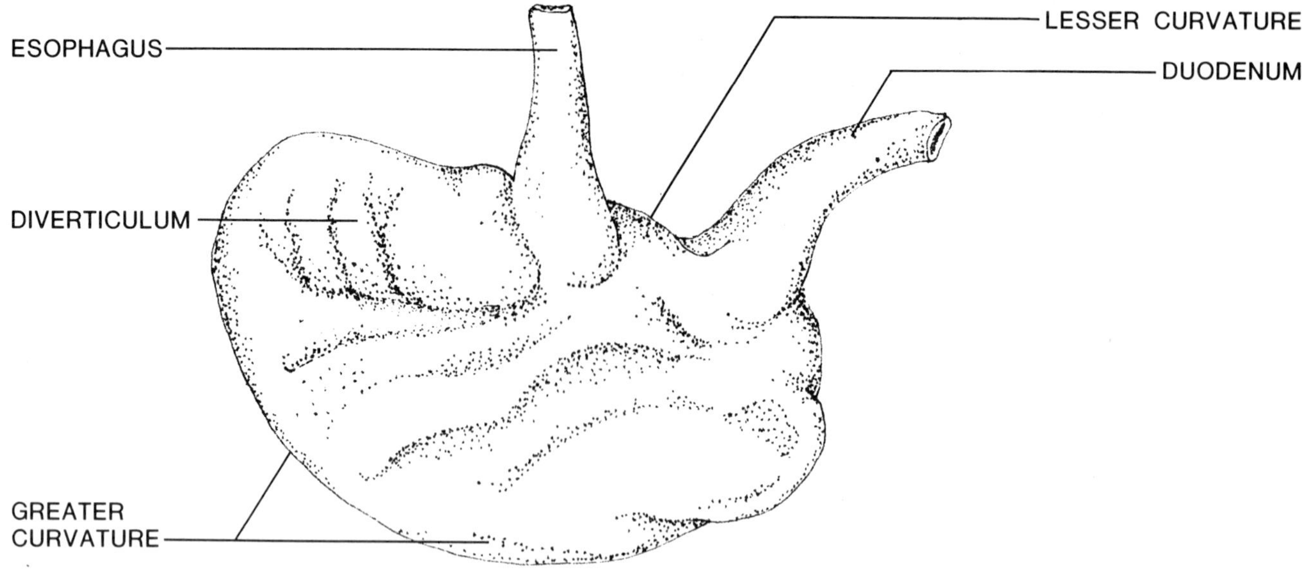

FIGURE 5.5a. Stomach, external (dorsal) view.

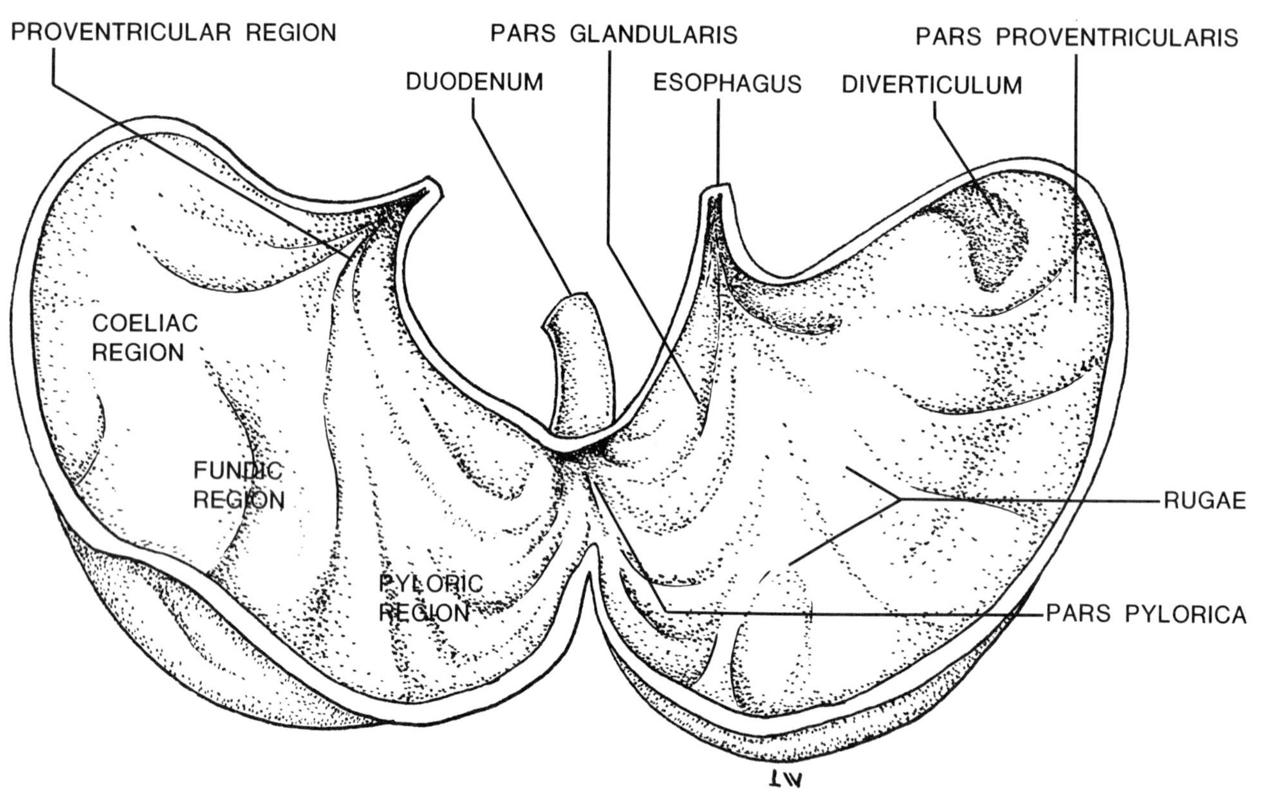

FIGURE 5.5b. Stomach, internal view.

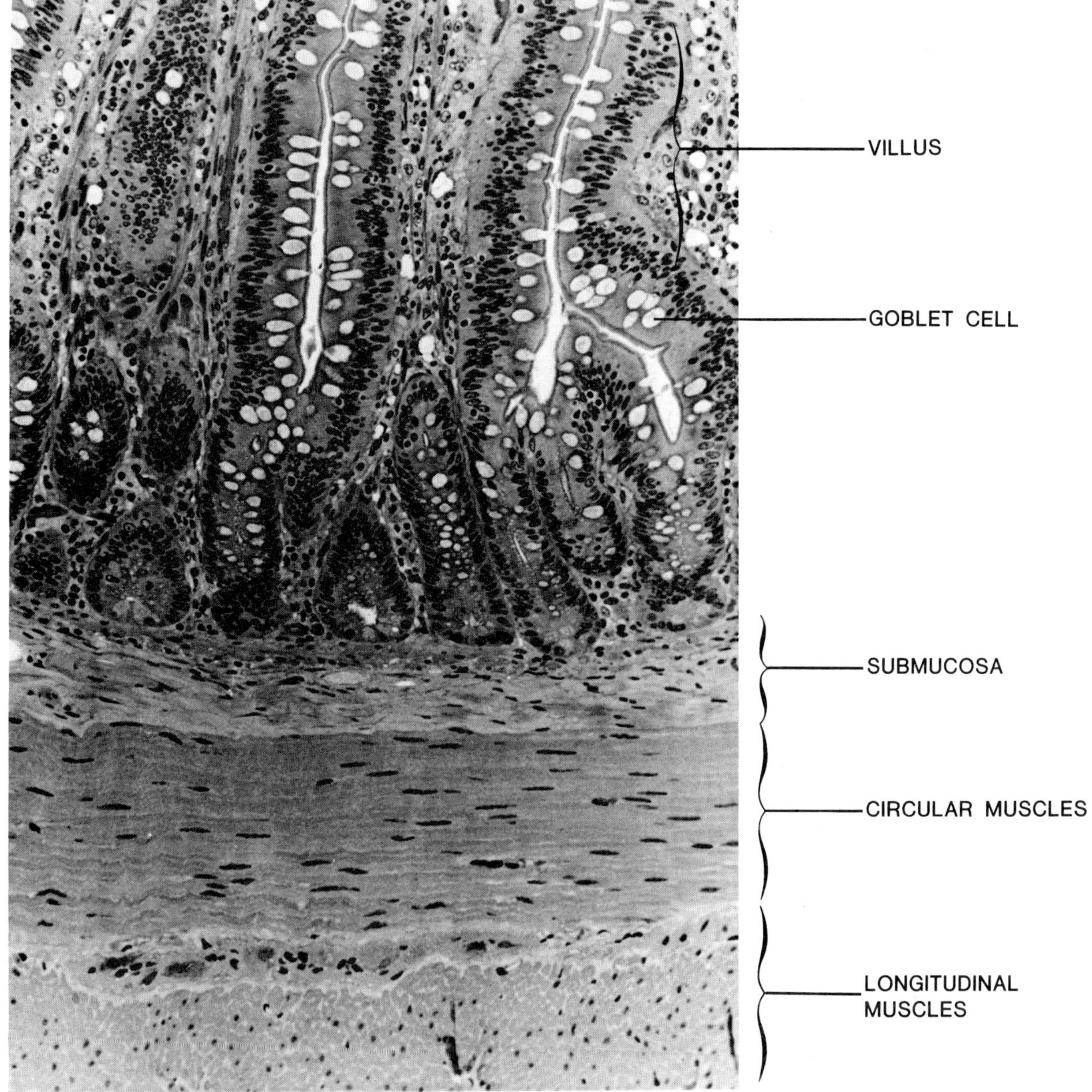

FIGURE 5.6. Cross section of portion of ileum (photomicrograph ×190).

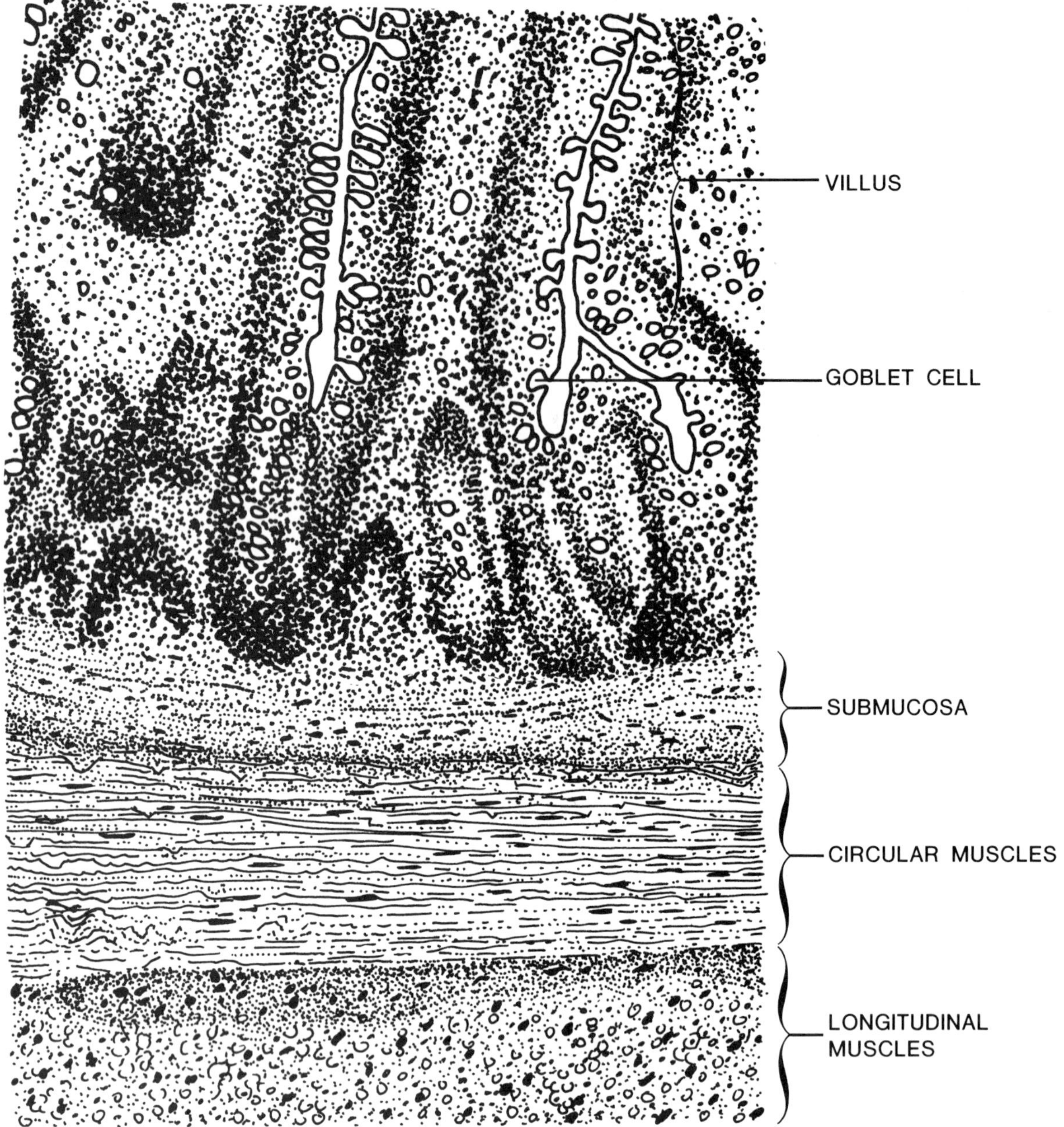

FIGURE 5.7. **Cross section of portion of ileum (diagrammatic).**

Chapter 6
Urogenital System

(Figures 6.1, 6.2, 6.3, 6.4, 6.5, 6.6, 6.7, 6.8, 6.9, 6.10, 6.11, 6.12, 6.13, 6.14, 6.15, 6.16)

CHAPTER OBJECTIVES

1. To identify the organs of the urinary system and those of the genital system.
2. To identify the microscopic structure of the kidney, ovary, and testis.

INTRODUCTION

In the fetal pig, the primary excretory organs are a pair of kidneys. Nitrogenous waste products from the blood are filtered through microscopic renal corpuscles in the kidneys, carried through a series of tubules and ducts to a bladder and eventually voided from the body through the urethra.

In the male, the urethra serves a dual function, that of carrying urine and sperm. In the female, the urethra joins the vagina to form a common duct called the urogenital sinus before exiting from the body.

EXCRETORY SYSTEM

A pair of kidneys is situated in the lumbar region against the dorsal body wall. They are covered by the *peritoneum,* the smooth shiny membrane that lines the body cavity or *coelom;* this must be removed before the kidneys can be clearly seen. Emerging from the medial face of the kidney is the *ureter* which empties into the *urinary bladder.* The bladder is an elongated sac lying between two large blood vessels, the umbilical arteries. The bladder continues into the umbilical cord where it is known as the *allantoic duct;* this duct will close following the birth of the fetus.

With a scalpel, cut through the cartilage and muscles of the pelvic girdle and lay the legs out flat. Dissect as necessary to expose the *urethra,* proceeding caudally from the urinary bladder. Follow the urethra to its continuation with the penis in the male and to its union with the vagina in the female.

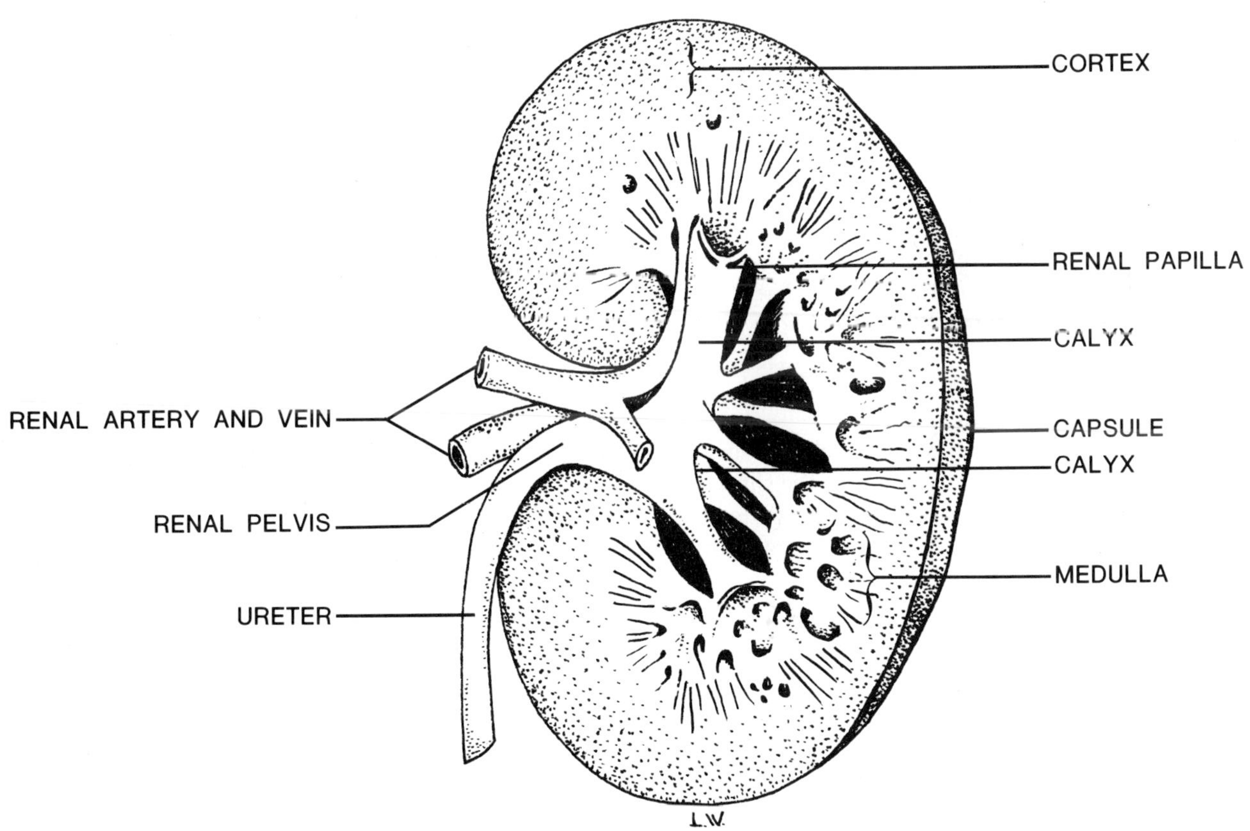

FIGURE 6.1. Sagittal section of kidney.

Look on the medial face of the kidney near its cranial end for the *suprarenal* or *adrenal* gland, part of the *endocrine* system. *Hormones* secreted by the adrenal glands directly into the bloodstream help to control a number of bodily functions such as heartbeat, regulation of blood sugar levels, dilation of muscles of bronchioles, and development of certain secondary sexual characteristics.

Remove one of the kidneys and a portion of its ureter from the body. The concave medial face of the kidney is the *hilum.* Cut a sagittal section through the kidney and study its cut surface. Note the *renal pelvis,* the expanded portion of the ureter within the kidney, and the smaller subdivisions of the pelvis, which are the *calyces.* The central part of the kidney containing blood vessels, pelvis, and calyces is called the *medulla.* Surrounding the medulla is the area known as the *cortex.*

FEMALE REPRODUCTIVE SYSTEM
(Figures 6.6, 6.7, 6.8, 6.9)

1. *Ovaries.* Small bean-shaped bodies lying on the dorsal body wall near the caudal end of the kidneys. Each one is suspended by a sheet of connective tissue called the *broad ligament.* A second mesentery, the *round ligament,* crosses the broad ligament diagonally from its upper edge and attaches to the caudal wall of the pelvic cavity.
2. *Uterine tubes.* Small compact coiled tubes on the dorsal face of each ovary. The opening from the body cavity into each tube is the *ostium* which is surrounded by small projections called *fimbriae.*
3. *Uterine horns* or *cornua.* These are the larger convoluted continuations of the uterine tubes attached to the free edge of the broad ligament. The two horns unite into a common structure, the *body* of the uterus. This is the *bicornuate* type of uterus in which the young develop in the horns and not within the body of the uterus.
4. *Vagina.* A muscular tube directly continuous with the body of the uterus.
5. *Urogenital sinus.* A common area into which both urethra and vagina open.
6. *Vulva.* This includes the external genitalia such as the *labia,* or lips, on either side of the slitlike urogenital opening and the *clitoris,* a small rounded body of erectile tissue on the ventral floor of the urogenital sinus.

MALE REPRODUCTIVE SYSTEM
(Figures 6.10, 6.11, 6.12)

1. *Scrotal sac.* Slit the scrotal sac in the midline and pull out the two dark-colored sacs. These sacs are *vaginal tunics* and enclose the *testes.* In order to expose the testes, the sacs must be removed.

2. *Gubernaculum.* The tough white cord that connects the caudal end of the testis to the inner face of the sac. The gubernaculum is homologous to the round ligament in the female reproductive system.
3. *Epididymis.* A tightly coiled mass of tubules consisting of a head, body, and tail and lying along one side of the testis from its cranial to its caudal end. The epididymis connects the *seminiferous tubules* of the testis with the sperm duct or ductus deferens.
4. *Ductus deferens.* Proceeds from the tail of the epididymis toward the abdominal wall passing through the inguinal canal into the abdominal cavity and empties into the urethra.
5. *Spermatic cord.* Consists of the ductus deferens, spermatic artery, spermatic vein, and spermatic nerve proceeding together from the testis through the inguinal canal.
6. *Inguinal canal.* An "opening" surrounded by an abdominal ring of muscle between the abdominal cavity and the cavity of the scrotal sac. It is through this ring that the testes descend from the abdominal cavity during their development into the scrotum. After emerging into the abdominal cavity, the ductus deferens turns medially, loops over the ureter, and enters the dorsal surface of the urethra close to the entrance of the ductus deferens from the opposite side. The spermatic artery joins the abdominal aorta, the spermatic vein empties into the caudal vena cava, and the spermatic nerve arises from the caudal spinal cord.
7. *Penis.* A white tubular structure that was seen previously when the skin and connective tissue were removed from the midline of the body caudal to the umbilical cord. In order to expose the full length of the penis and to identify the urethra and its associated glands, it will be necessary to make the following dissection. Make an incision through the muscles in the midventral line between the hindlegs. Continue the cut through the pubic symphysis until the hindlegs lie flat. Remove enough muscle and pubic bone on each side of the cut until the urethra is exposed. Free the urethra from the rectum which lies dorsal to it by cutting through any muscles or connective tissue you may encounter.
8. *Seminal vesicles.* A pair of small glands situated on the dorsal surface of the urethra at the point where the ductus deferens from each side enters.
9. *Prostate gland.* A small gland located between the bases of the seminal vesicles. It may be difficult to locate because of its small size and immaturity. The prostate shown in figure 6.12 is

diagrammatic and is included only to illustrate the complete male reproductive system.

10. *Bulbourethral glands.* Two large elongated glands lying on either side of the junction between the urethra and penis.

MICROSCOPIC ANATOMY OF KIDNEY, OVARY, AND TESTIS

(Figures 6.2, 6.3, 6.4, 6.13, 6.14, 6.15, 6.16)

Prepared microscope slides of the sections of adult organs will show most of the following structures.

1. *Kidney.* The kidney consists of two major regions, the *cortex* and the *medulla,* the whole organ being enclosed in a connective tissue sheath, the *capsule.* Within the cortex are the numerous *renal corpuscles* which filter the waste products of metabolism from the blood. The structure of a corpuscle is seen in figure 6.3. Collecting tubules, called *medullary rays,* extend from the cortex into the medulla. The urine in these smaller tubules empties into larger collecting tubules, then to a common collecting area, the *pelvis,* and finally into the ureter and the urinary bladder.

2. *Ovary.* The *cortex* or outer portion contains a number of *follicles* in varying stages of development. It is within the follicle that the *ovum* or egg is formed. Figure 6.13 shows a mature follicle enclosing a small mass of cells, the *cumulus oophorus* or "egg hill" in which the nucleus of the ovum may be seen. The large open area, the *antrum,* is filled with a fluid called the *liquor folliculi.* A female sex hormone, *estrogen,* is produced by the follicle and is associated with the secondary sexual characteristics of the female.

3. *Testis.* This is a compound tubular gland composed of a fibrous connective tissue capsule enclosing large numbers of *seminiferous tubules.* The area surrounding the lumen or space within each tubule is the site of development of the *spermatozoa.* These begin as *spermatogonia* near the periphery of the tubule, undergoing maturation as they approach the lumen of the tubule. Associated with the developing sperm are support and nutrient cells called *Sertoli* cells. Lying between the seminiferous tubules are specialized cells which have an endocrine function. These are the *interstitial* cells that produce the male sex hormone, *testosterone,* which is responsible for the development and maintenance of the secondary sexual characteristics of the male.

4. *Penis.* Use a sharp scalpel or razor blade and cut out a thin cross-sectional slice from the penis. Place this on a slide with a drop or two of water and examine with the dissecting microscope. Observe the two honey-combed areas on the dorso-lateral side; these are the *corpora cavernosa penis.* On the ventral side of the penis is a single mass of tissue, the *corpus spongiosum urethrae;* within this is a small slit, the *urethral canal,* which serves to conduct sperm and urine. The three corpora serve as erectile tissue for the penis.

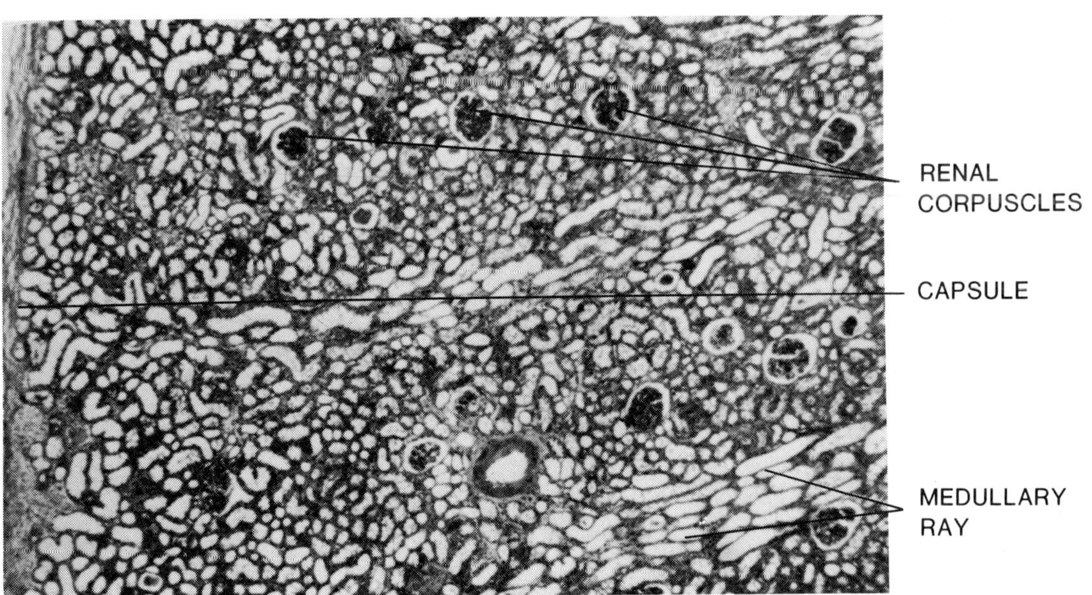

FIGURE 6.2.   Section of cortex of kidney (photomicrograph ×10).

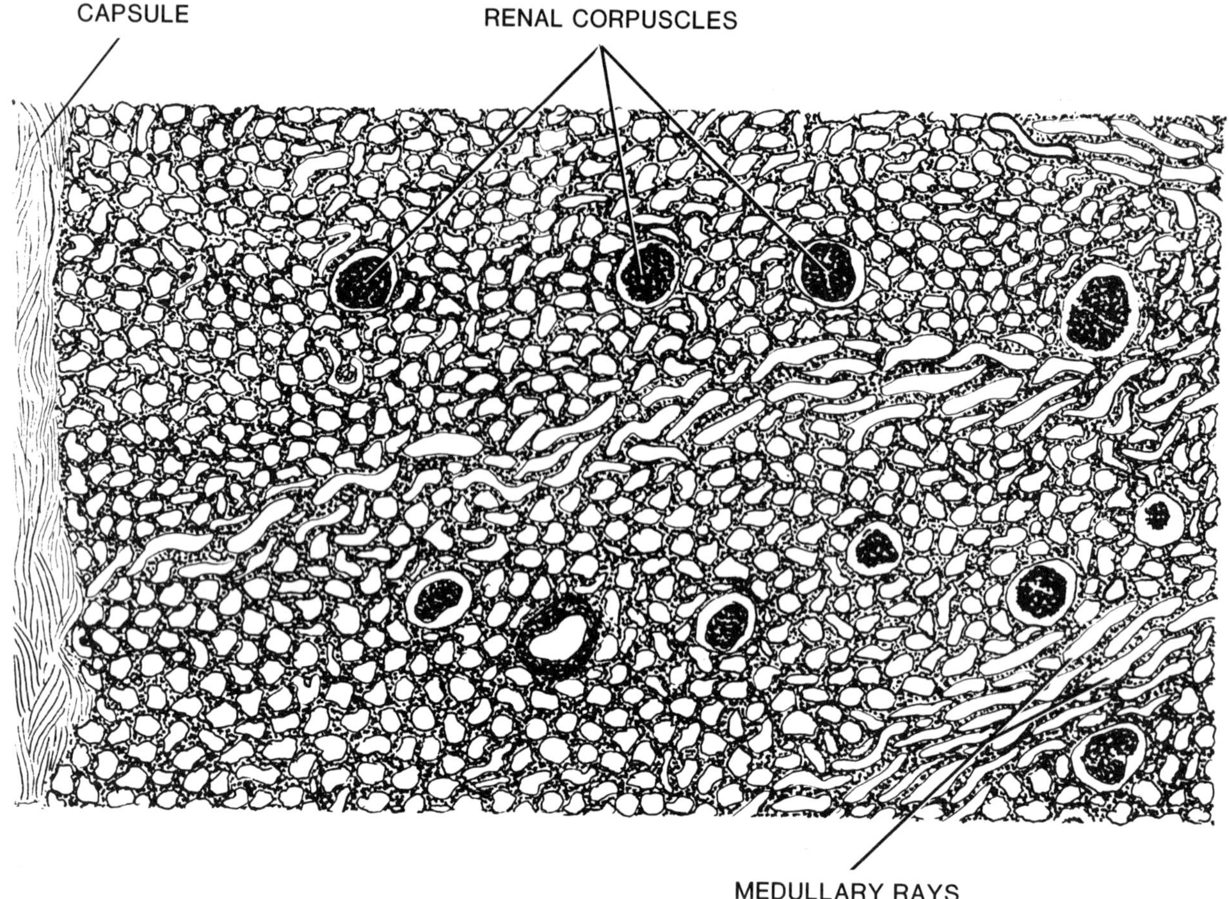

FIGURE 6.3.   Section of cortex of kidney (diagrammatic).

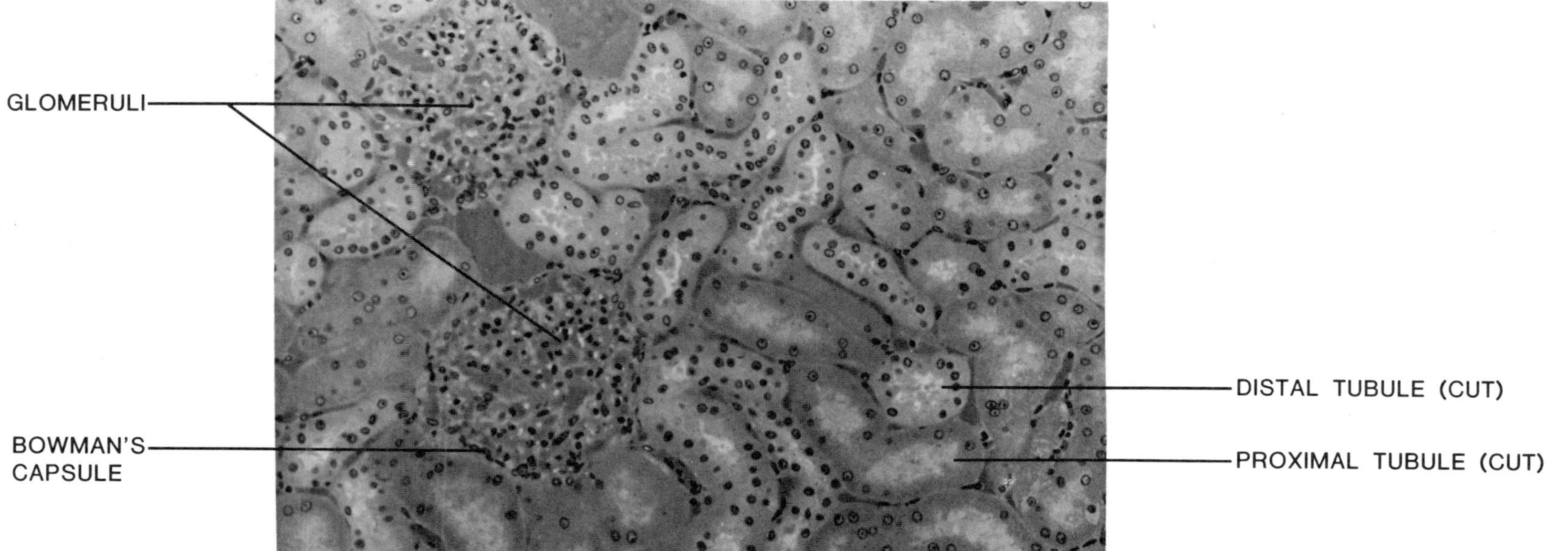

FIGURE 6.4. Section of cortex of kidney (photomicrograph ×480).

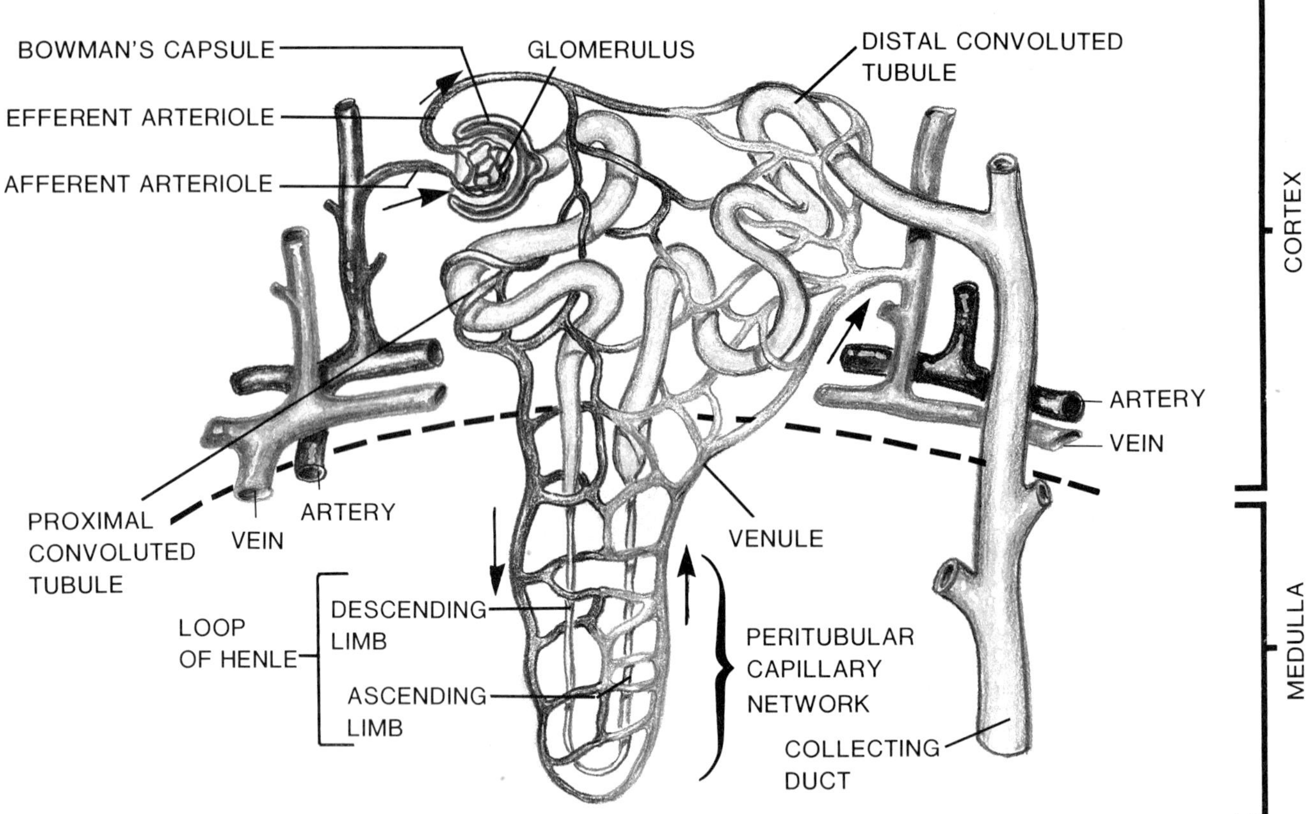

FIGURE 6.5. **The mammalian nephron (diagrammatic).**
From John W. Hole, Jr., *Human Anatomy and Physiology*, 5th ed.
Copyright © 1990 Wm. C. Brown Publishers, Dubuque, Iowa. All
Rights Reserved. Reprinted by permission.

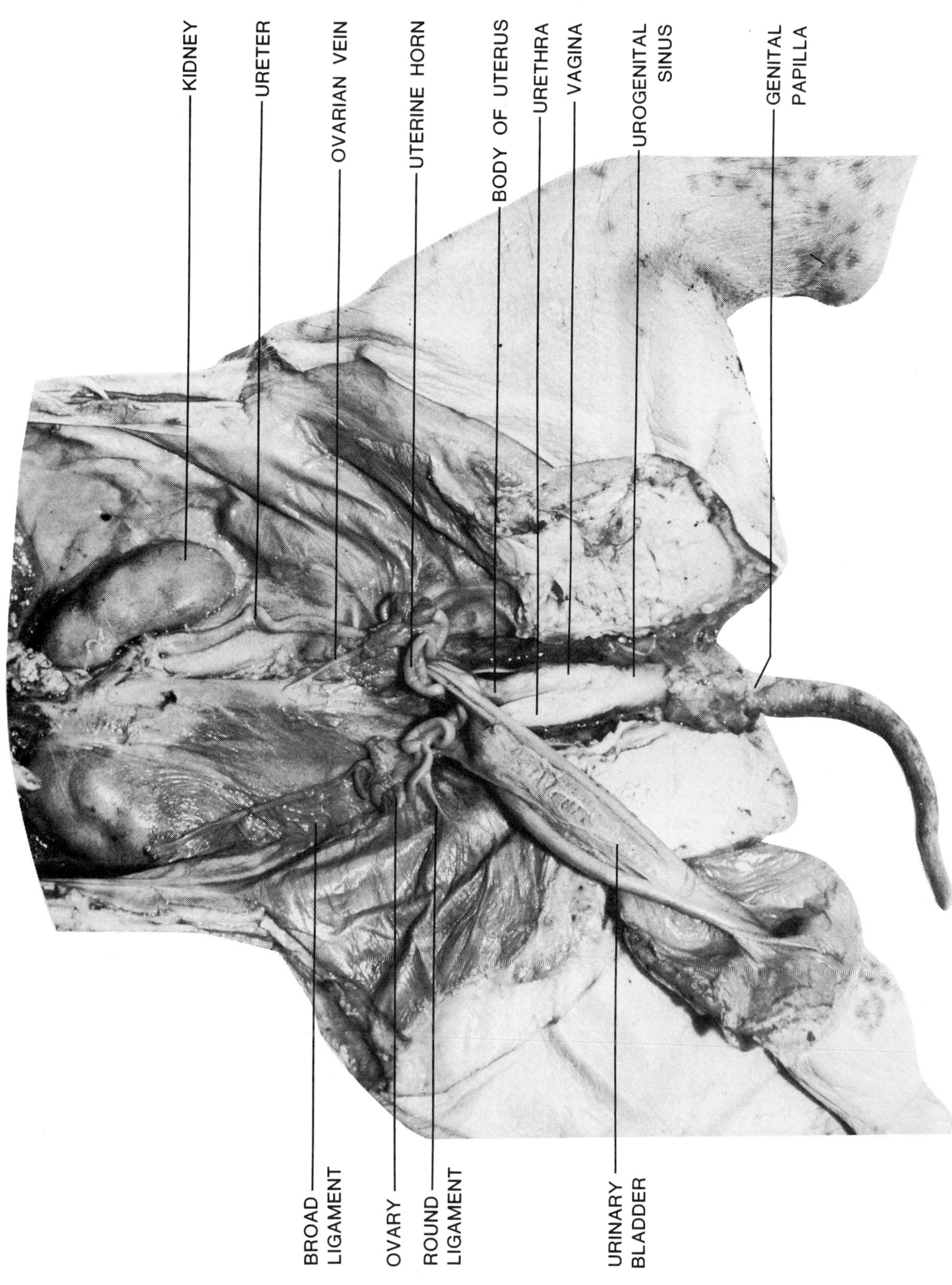

FIGURE 6.6. Female urogenital system, ventral view.

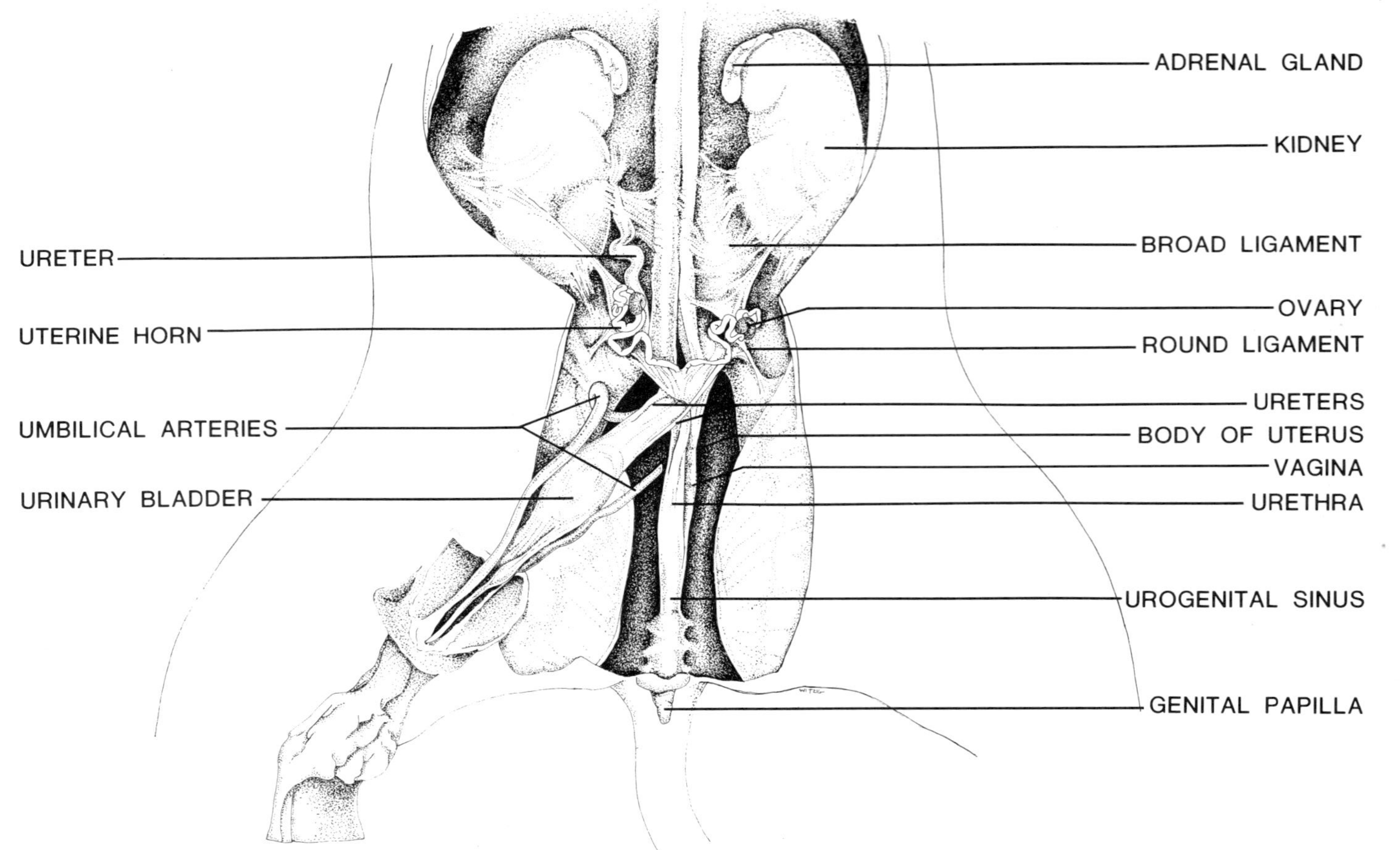

FIGURE 6.7. Female urogenital system, ventral view.

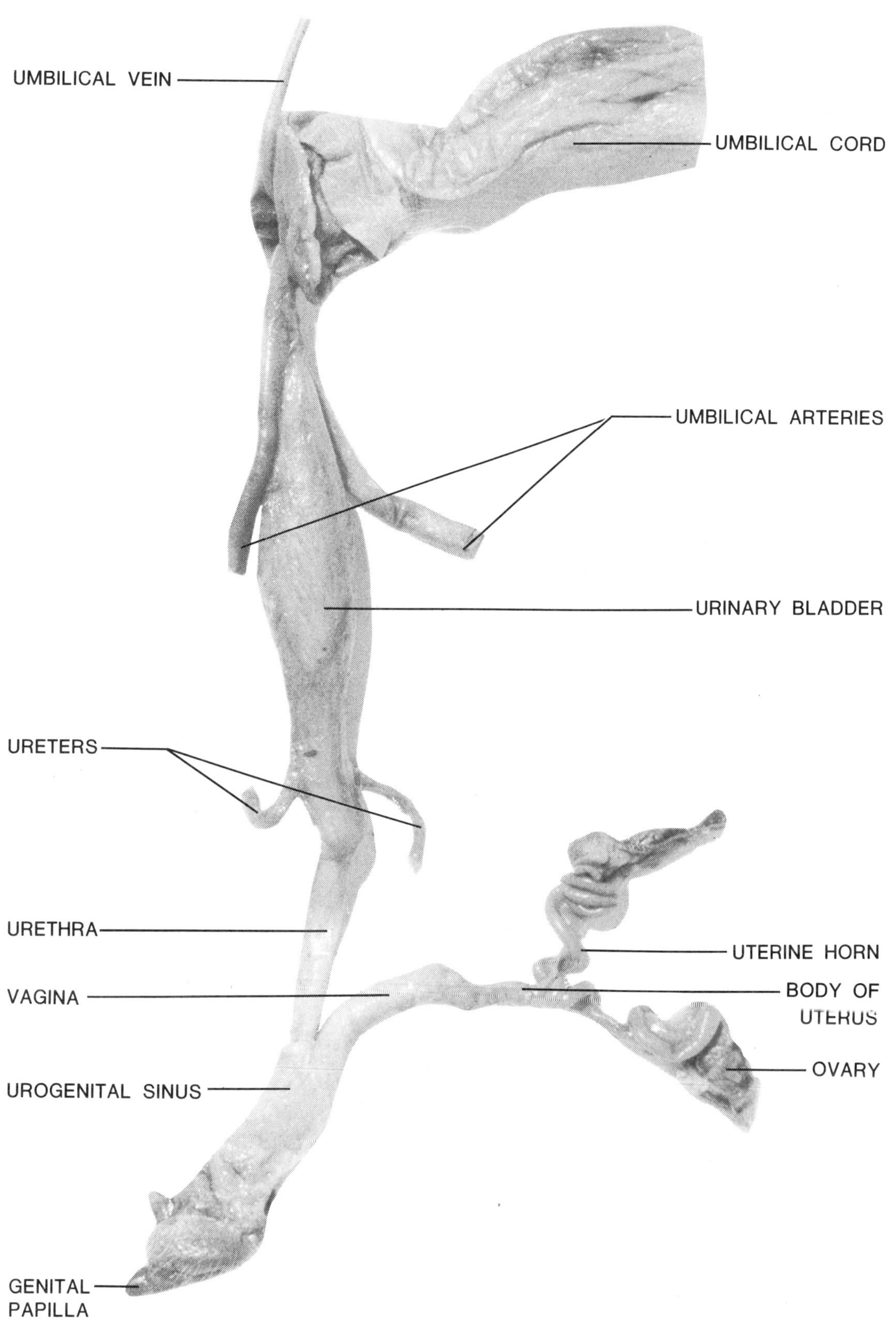

FIGURE 6.8. Female urogenital system, dorsal view.

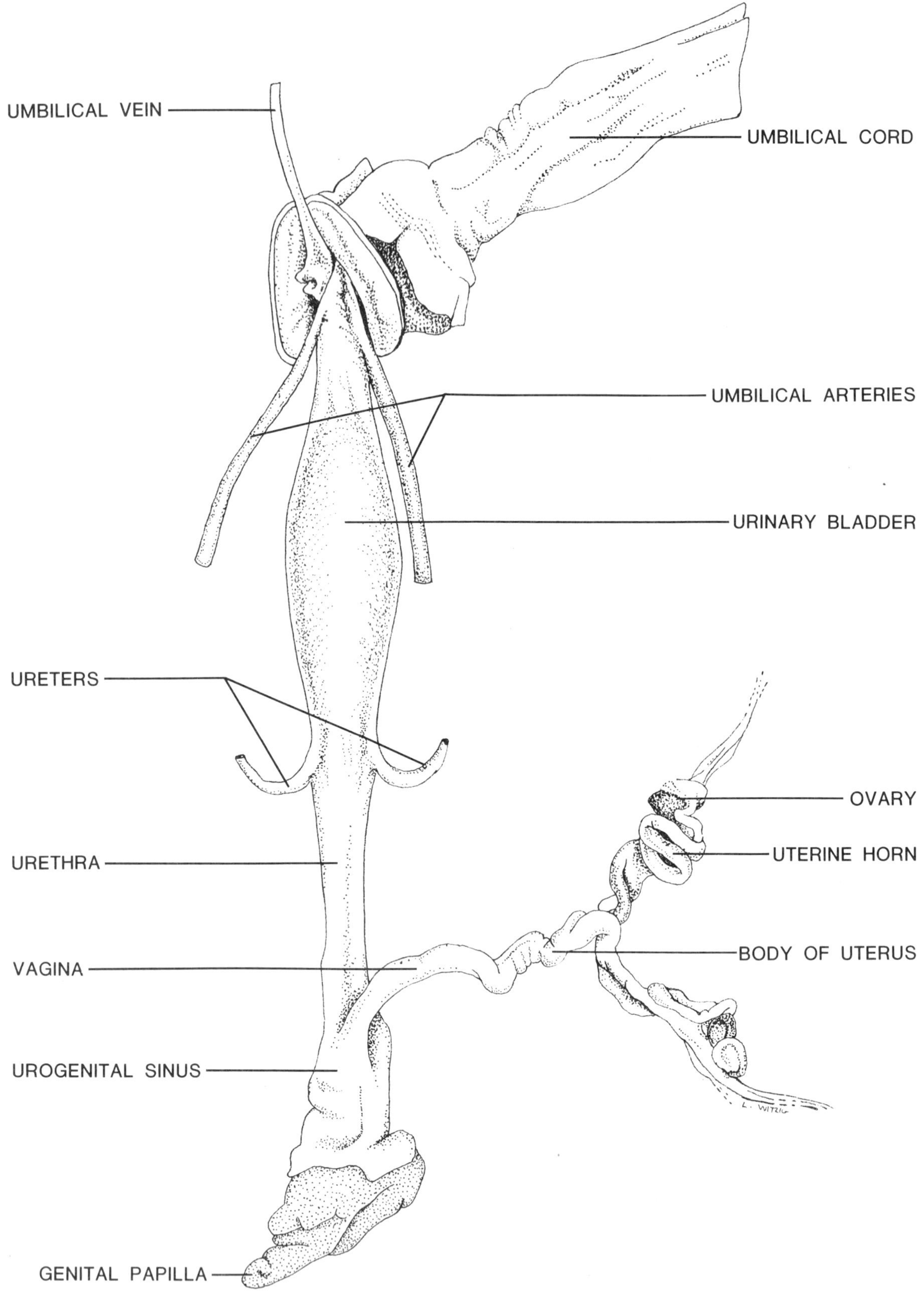

FIGURE 6.9. Female urogenital system, dorsal view.

52

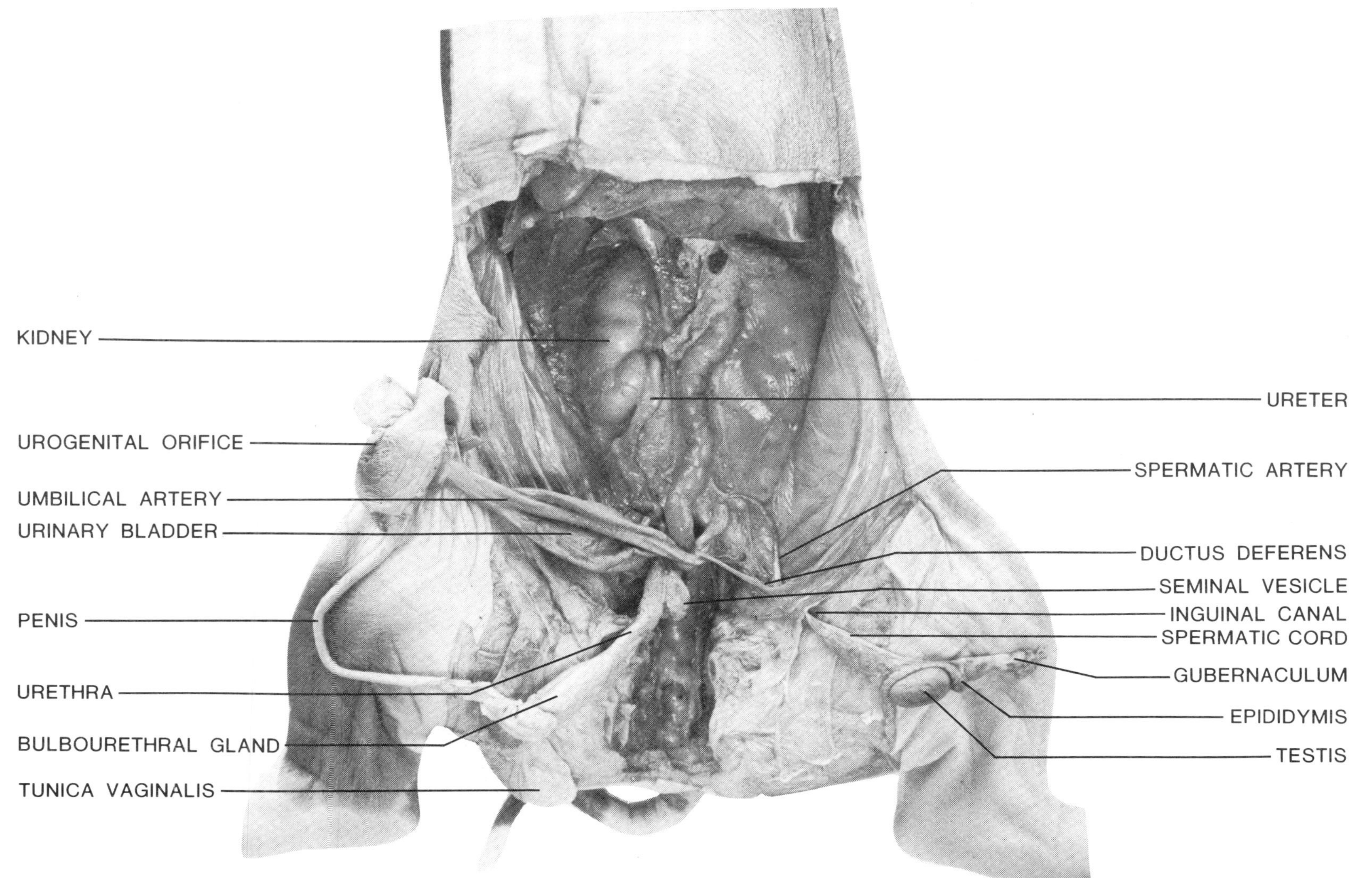

FIGURE 6.10.   Male urogenital system, ventral view.

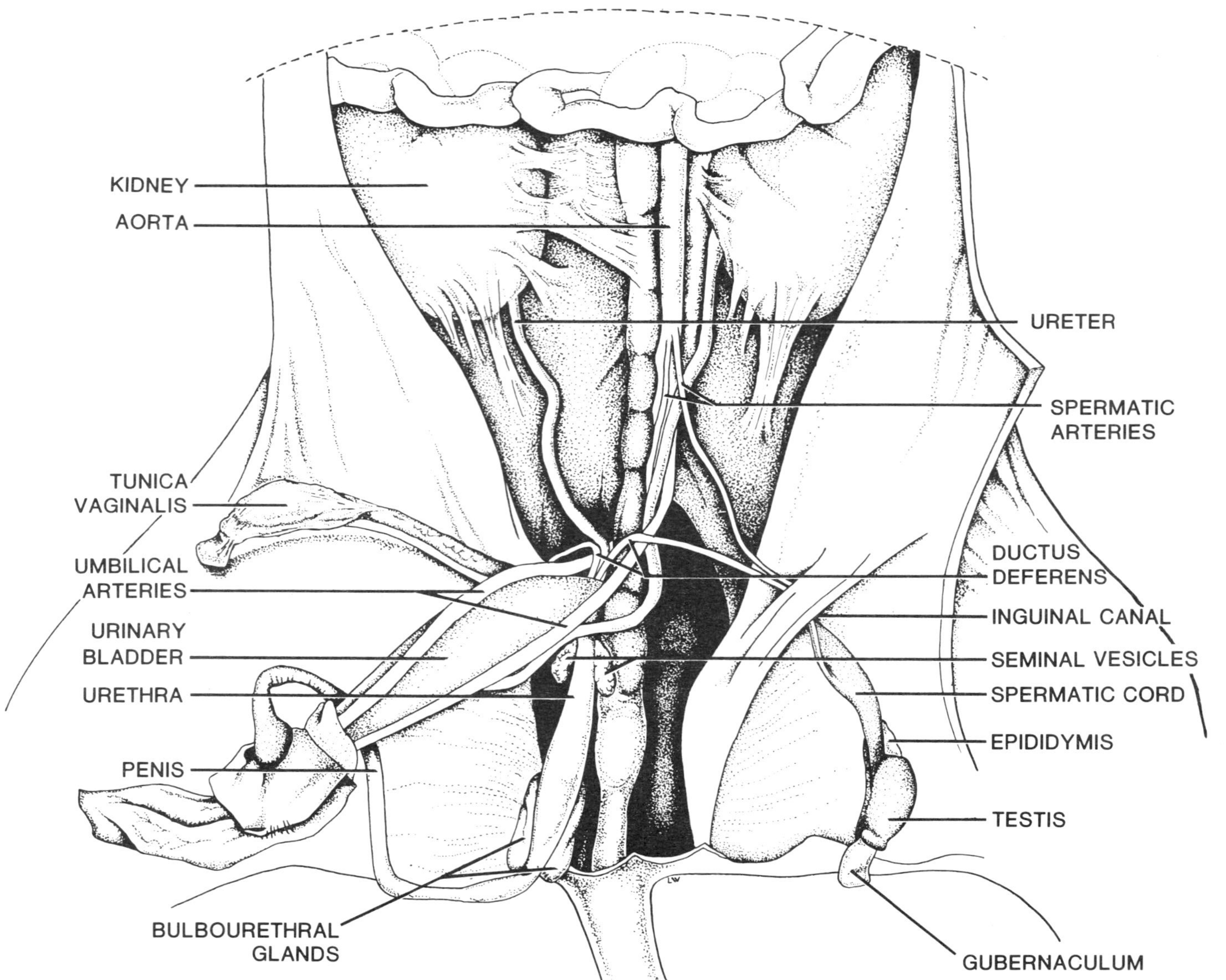

FIGURE 6.11. Male urogenital system, ventral view.

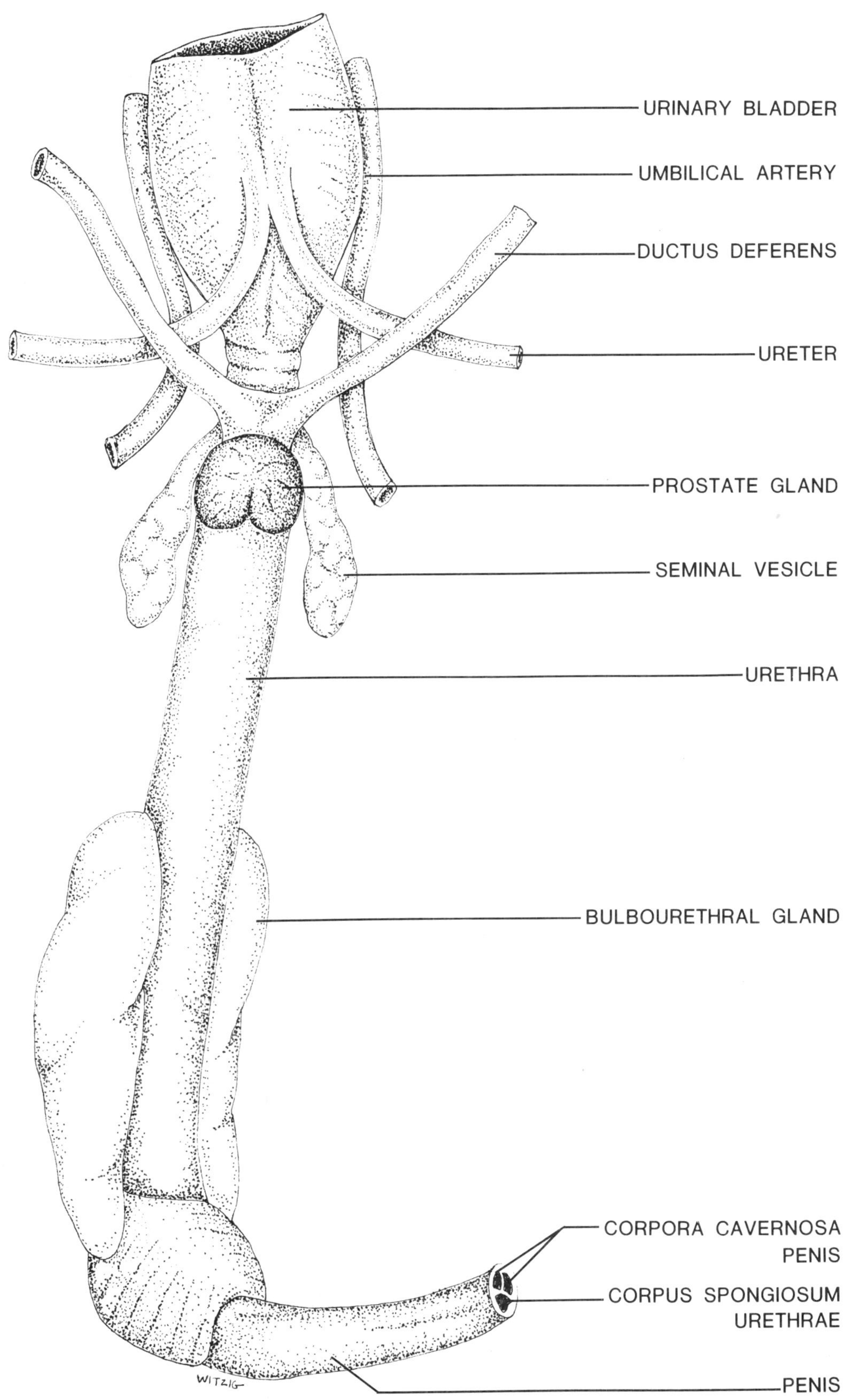

FIGURE 6.12. **Male urogenital system, dorsal view.**

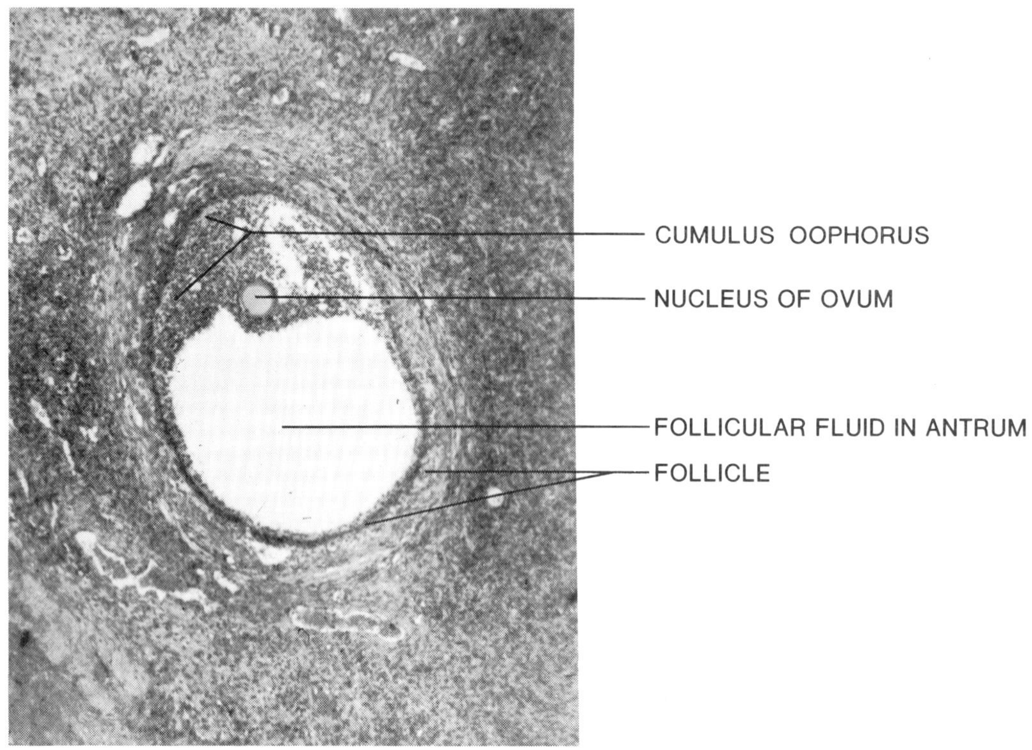

FIGURE 6.13. Section of ovary showing mature follicle (photomicrograph ×75).

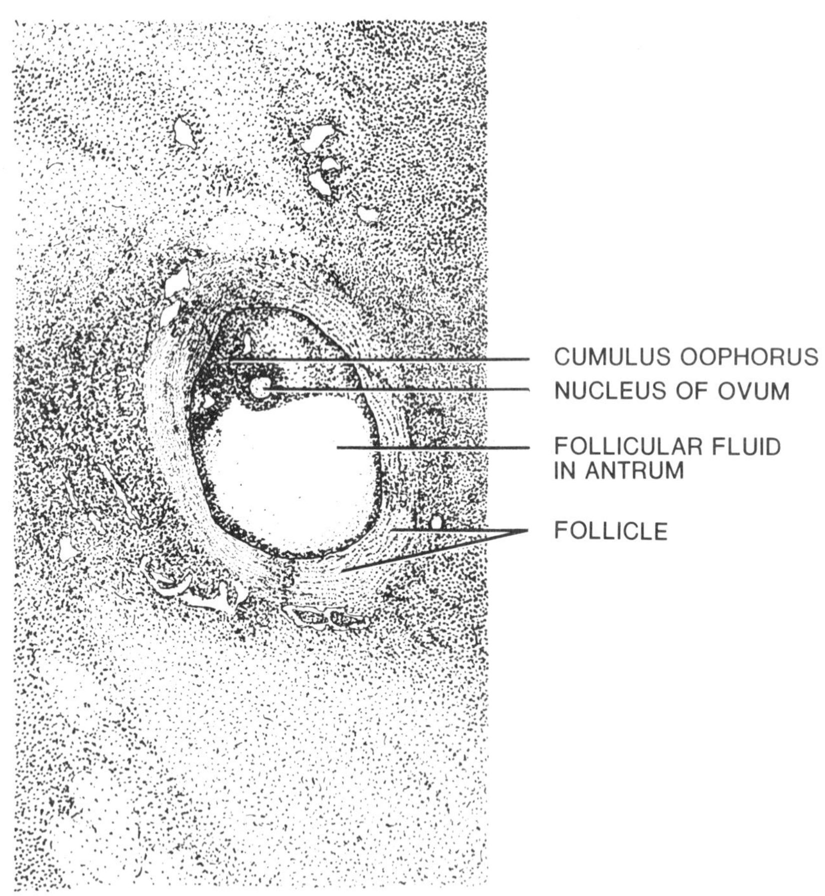

FIGURE 6.14. Section of ovary showing mature follicle (diagrammatic).

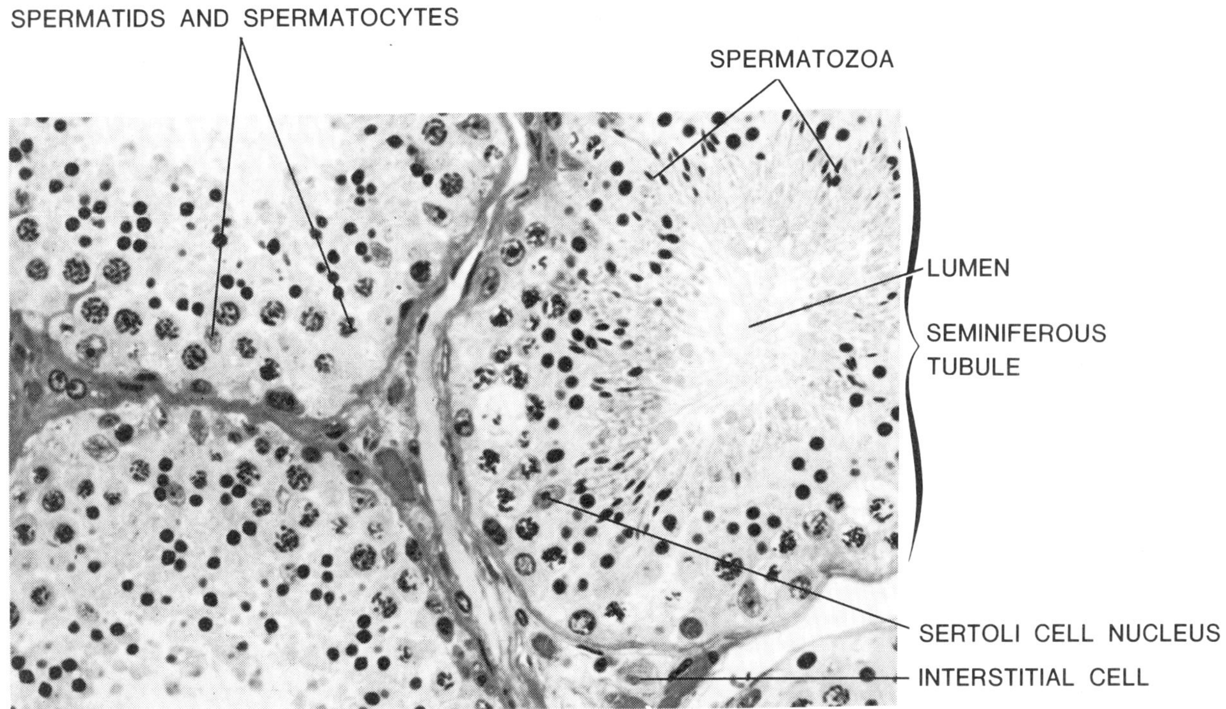

FIGURE 6.15. Section of testis (photomicrograph ×350).

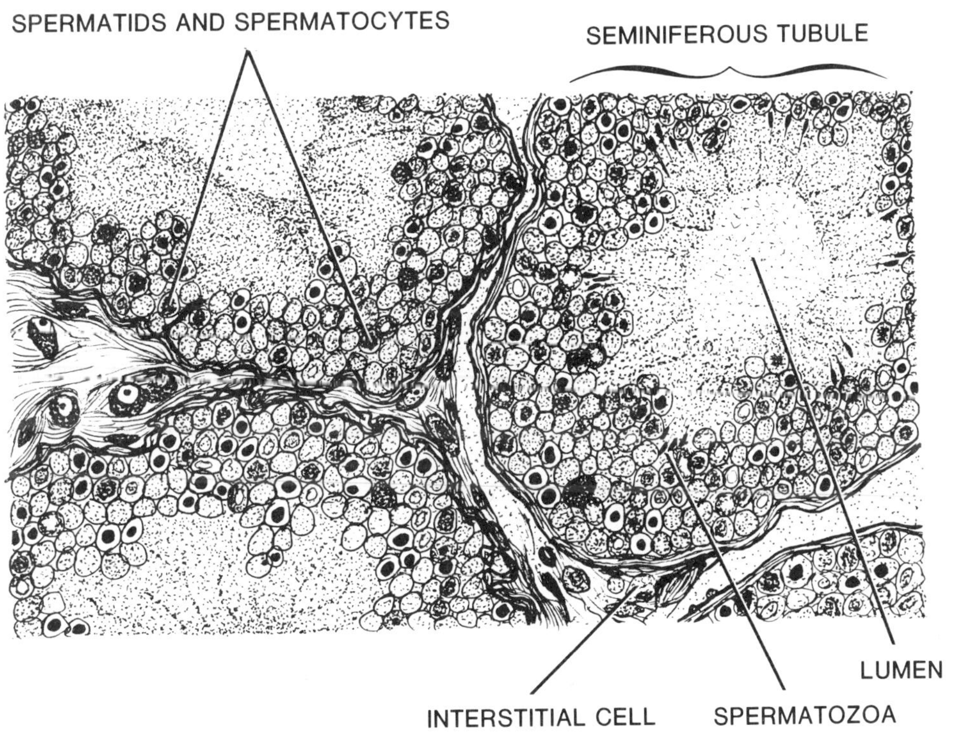

FIGURE 6.16. Section of testis showing seminiferous tubules (diagrammatic).

Chapter 7
Circulatory System

(Figures 7.1, 7.2, 7.3, 7.4, 7.5a, 7.5b, 7.6, 7.7, 7.8, 7.9, 7.10, 7.11, 7.12)

CHAPTER OBJECTIVES

1. To identify the structures of the heart.
2. To identify the major arteries and veins and to trace the course of blood through these vessels.
3. To compare fetal circulation with circulation in the adult.

INTRODUCTION

The circulatory system provides the means, through the plasma and cellular components of the blood, by which carbon dioxide and nitrogenous wastes are transported to their respective exits from the body.

In the adult, arteries carry blood from the heart to the various organs of the body while veins return blood from these organs back to the heart. With one exception, blood in the arteries is oxygenated whereas blood in the veins is not. The exception is the pulmonary artery, which carries nonoxygenated blood from the right ventricle of the heart to the lungs; the pulmonary veins return oxygenated blood from the lungs to the left atrium. The condition in the fetus is different since the lungs are not functional and the right and left atria are interconnected. (See page 70 for a comparison of fetal and adult circulation.)

The circulatory system consists of the *heart;* the *arteries,* which carry blood away from the heart; the *veins,* which carry blood to the heart; and the *lymphatic vessels* and *nodes.* The lymphatics collect *tissue fluid* produced by diffusion and filtration of blood *plasma* at the capillary ends of the arteries and return it to the venous system cranial to the heart.

The drawings of the circulatory system show the arteries in red and the veins in blue—this represents the situation as the student views the doubly injected specimens. However, the student should not accept that the color red indicates vessels carrying blood rich in oxygen or that blue indicates vessels deficient in oxygen. In fetal circulation, there is considerable mixing of the two kinds of blood. The only vessel transporting relatively oxygen-rich blood is the umbilical vein from the placenta. Before this blood can reach the right atrium of the heart it is emptied into the caudal vena cava carrying oxygen-deficient blood from the caudal part of the body. This mixed blood in the right atrium is further mixed with oxygen-deficient blood entering the right atrium from the cranial vena cava.

Some of the mixed blood in the right atrium passes through the foramen ovale to the left atrium, then to the left ventricle, and thus to the aorta. Most of the blood in the right atrium is pumped into the right ventricle. In the adult, this blood is directed into the lungs via the pulmonary artery but in the fetus it is bypassed into the aorta through the ductus arteriosus. From the aorta, blood is carried to all parts of the fetal body including the flow through the umbilical arteries to the placenta where the blood will again be oxygenated.

HEART

Before beginning the following dissection, remove the thymus gland, thyroid, and muscles in the throat region.

Remove the pericardial sac, which surrounds the heart, and identify the four chambers. These consist of thin-walled right and left *atria* whose position may be approximated by observing the lobed or scalloped appendages, the *auricles,* lying on the atria, and the large right and left *ventricles.* Lying in the diagonal groove between the ventricles are the *coronary* artery and vein.

Proceeding out of the right ventricle and lying between the two auricles is a large vessel, the *pulmonary trunk,* which, in the adult, carries blood to the lungs to be oxygenated. Follow this vessel and identify its two branches, the right and left pulmonary *arteries* to the lungs.

The two large dark vessels entering the right atrium are the *cranial* and *caudal venae cavae.*

The large vessel whose base is covered by the pulmonary trunk is the aorta, which leads out of the left ventricle. The aorta turns almost immediately caudally and continues through the thoracic and abdominal cavities. The point where it turns caudally is the *arch* of the aorta. Arising from the arch are two major vessels, one of which is the *brachiocephalic* or *innominate* artery, which will give rise to the *carotid* trunk to the head and the *right subclavian* artery to the foreleg. The other major branch from the arch is the *left subclavian* artery to the left foreleg.

In the fetus there is a short connection between the pulmonary trunk and the aorta. This is the *ductus arteriosus.* Since the fetal lungs are not functional, blood from the right ventricle is shunted into the aorta rather than flowing to the lungs. After the birth of the fetus, the ductus arteriosus becomes a functionless solid cord, the *arterial ligament.*

Pulmonary veins from the lungs to the left atrium can be seen by scraping away some of the lung tissue where the lungs lie close to the heart.

The internal structures of the heart will be studied following an examination of the arteries and veins.

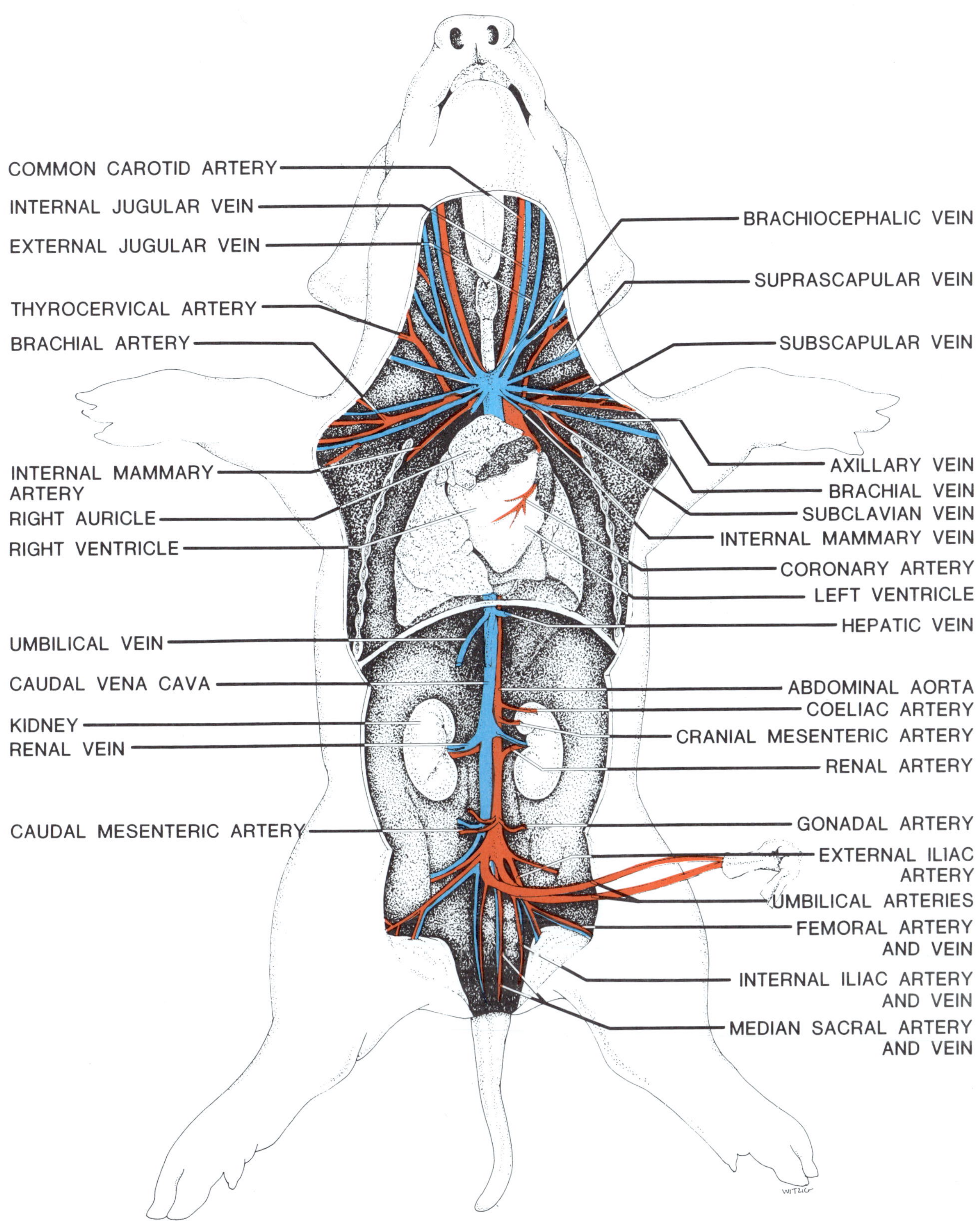

FIGURE 7.1. Arteries and veins, ventral view.

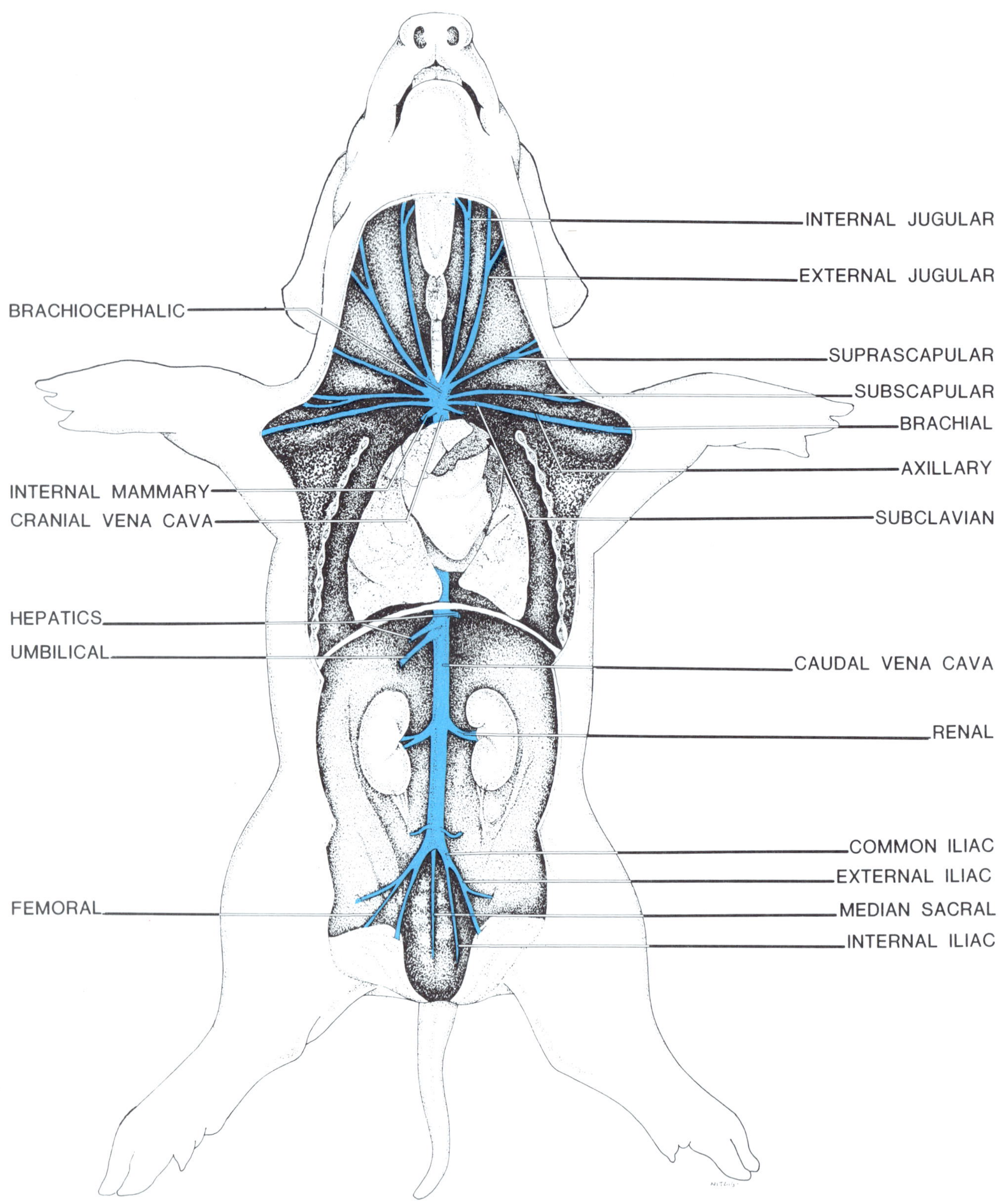

FIGURE 7.2. Venous system, ventral view.

VEINS CRANIAL TO THE HEART
(Figures 7.1, 7.2)

These vessels collect blood from the head, neck, cranial appendages, and thorax and empty into the right atrium of the heart through the *cranial vena cava.*

1. *Cranial vena cava.* This is the large dark vessel that enters the right atrium. Trace it anteriorly and note that it is formed by the union of two short trunks, the *brachiocephalic* or *innominate* veins.
2. *External jugulars.* A pair of veins carrying blood from the head and face and emptying into the brachiocephalics.
3. *Internal jugulars.* Medial to the external jugulars, entering the external jugulars just before the latter empty into the brachiocephalics. The internal jugulars drain the vessels of the brain.

    There is considerable variability in the relationship of the jugular veins. Both internal and external jugulars may enter the brachiocephalic vein independently or the internal jugular may join with the external jugular before the latter empties into the brachiocephalic vein.
4. *Subclavians.* Well-defined vessels from the forelimbs entering the brachiocephalics at about the same point as the jugulars. Trace these vessels into the forelimb. Just external to the rib cage, the vessel is known as the *axillary* vein, in the upper portion of the limb as the *brachial,* in the lower foreleg as the *radial* and *ulnar* veins.
5. *Internal mammaries.* A pair of vessels arising independently or in common from the cranial vena cava at about the level of the first rib. These veins receive blood from the wall of the thorax. The connection of the internal mammary with the cranial vena cava may have been cut when you opened the chest cavity, but it can be seen by looking on the inner surface of the chest on either side of the midventral line.
6. *Hemiazygous.* Push the heart and lungs laterally and look for an unpaired vessel lying on the left side of the middorsal line. After receiving branches from the intercostal muscles, the vein continues toward the heart along the dorsolateral surface of the thoracic aorta, emptying into the right atrium through the coronary sinus in close proximity to the entrance of the caudal vena cava into the atrium.

VEINS CAUDAL TO THE HEART
(Figures 7.3, 7.4)

The *caudal vena cava* can be seen by lifting the heart anteriorly. The dark vessel entering the right atrium is the caudal vena cava. Trace the vessel caudally and identify the following veins that empty into it.

1. *Renal.* From the kidneys and frequently appearing as a split vessel, either on one side or both.
2. *Gonadals.* In the male, these are the *spermatic* veins and in the female, *ovarian* veins. These vessels usually enter the caudal vena cava caudal to the entrance of the renal veins, although frequently the gonadal vein on the left side empties into the left renal vein instead of the caudal vena cava.
3. *Common iliacs.* Paired vessels forming the major caudal division of the caudal vena cava.
4. *External iliacs.* The continuation of the common iliacs following the point of branching of the internal iliac. The external iliacs continue into the leg as the *femoral* veins.
5. *Internal iliacs.* Medial branches of the common iliac collecting blood from the pelvic viscera.
6. *Median sacral.* A small vein from the tail emptying into the caudal vena cava.
7. *Umbilical.* This is the vessel around which you tied a piece of string when the body cavity was first opened. The vessel carries oxygenated blood from the placenta, through the liver where the vessel is called the *ductus venosus,* to the caudal vena cava and the right atrium of the heart.
8. *Hepatic veins.* Scrape away the liver substance with the edge of the scalpel or forceps until only a small mass of cords remains. Identify the umbilical vein, ductus venosus, and hepatic veins. The latter collect blood from the liver and empty it into the caudal vena cava.

HEPATIC PORTAL SYSTEM
(Figure 7.3)

The hepatic portal system consists of veins that collect blood from the digestive tract and spleen and carry it to the liver. Here the portal vein divides into a network of capillaries. At the cranial end of the liver, the capillaries become hepatic veins that empty into the caudal vena cava. The vessels in this system are usually not injected but may contain sufficient blood to make it possible to trace the major ones.

1. *Portal vein.* Raise the liver lobes and look for the common bile duct lying in the *hepatoduodenal ligament.* The portal vein lies close to the duct and proceeds cranially to enter the liver.
2. *Gastrosplenic vein.* The first major branch of the portal vein to the left. This receives the *right gastroepiploic* vein from the pyloric region of the stomach and then continues as the *splenic* vein on the ventral side of the spleen. Approximately midway along its course the splenic vein is joined by the *left gastroepiploic* vein from the cardiac region of the stomach.

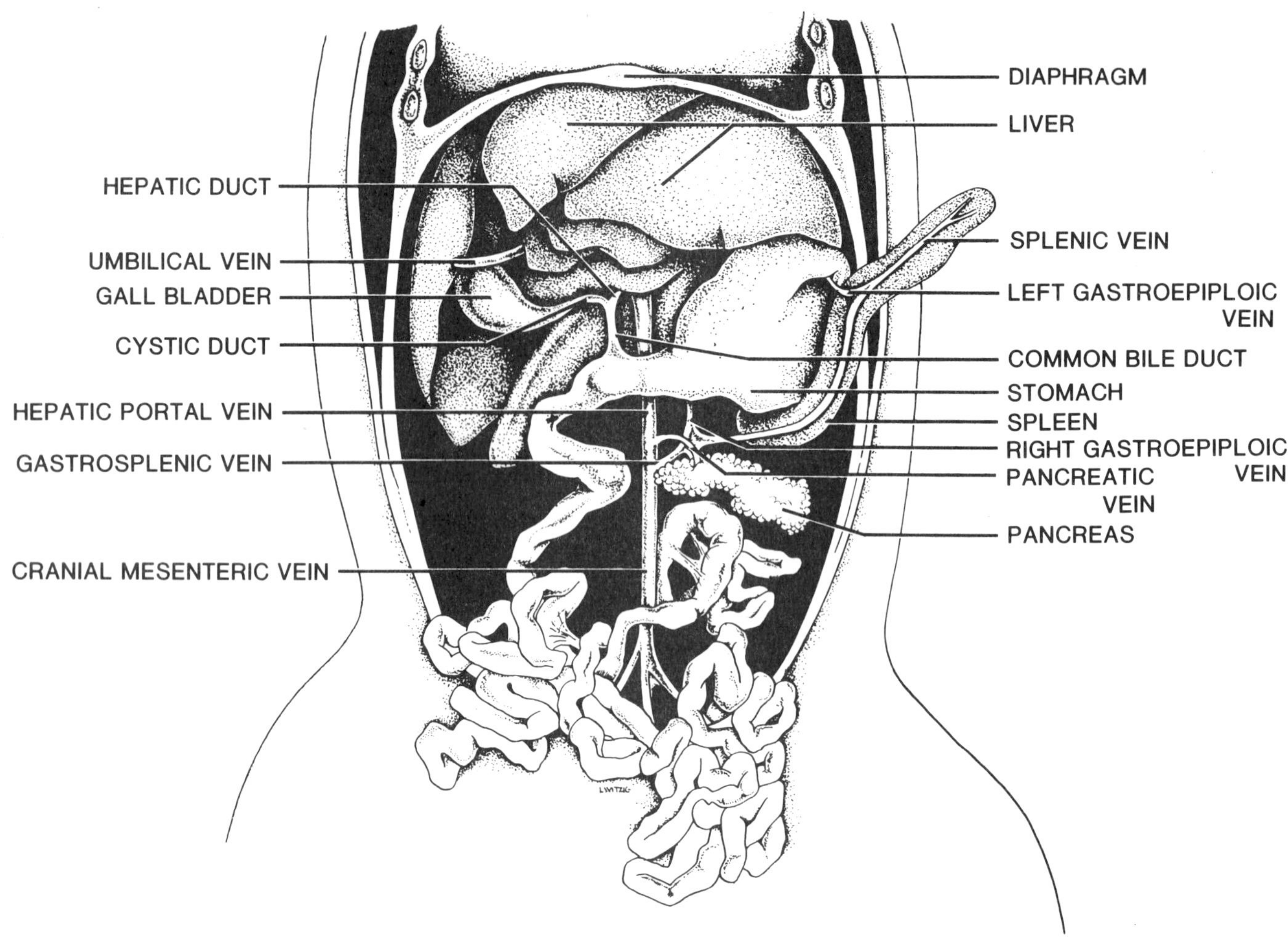

FIGURE 7.3. Hepatic portal system, gallbladder, bile ducts.

3. *Pancreatic vein.* A small vessel from the pancreas emptying directly into the portal vein just cranial to the entrance of the gastrosplenic vein.

4. *Cranial mesenteric vein.* This will appear as a direct continuation of the portal vein to the coils of the small intestine.

ARTERIES CRANIAL TO THE HEART
(Figures 7.1, 7.4, 7.5a, 7.5b, 7.6)

1. *Brachiocephalic.* A single large vessel from the arch of the aorta. The *carotid trunk* and the *right subclavian* artery are branches of this larger vessel.

2. *Carotid trunk.* A very short branch of the brachiocephalic, which gives rise to the two *common carotid* arteries.

3. *Common carotids.* Trace these forward as they lie on either side of the larynx. Look on the lateral surface of the carotids for a fine white fiber, the *vagus nerve.*

4. *External carotids.* Near the cranial end of the larynx, the common carotids branch into external carotids supplying the head and face.

5. *Internal carotids.* These branch from the common carotids at the same point as the external carotids to supply the brain with blood.

6. *Left subclavian.* This vessel arises directly from the arch of the aorta and proceeds laterally. Between the point of emergence from the thoracic cavity and the upper foreleg, the subclavian is called the *axillary* artery. It becomes the *brachial* artery in the upper foreleg, dividing in turn into the *radial* and *ulnar* arteries in the lower foreleg.

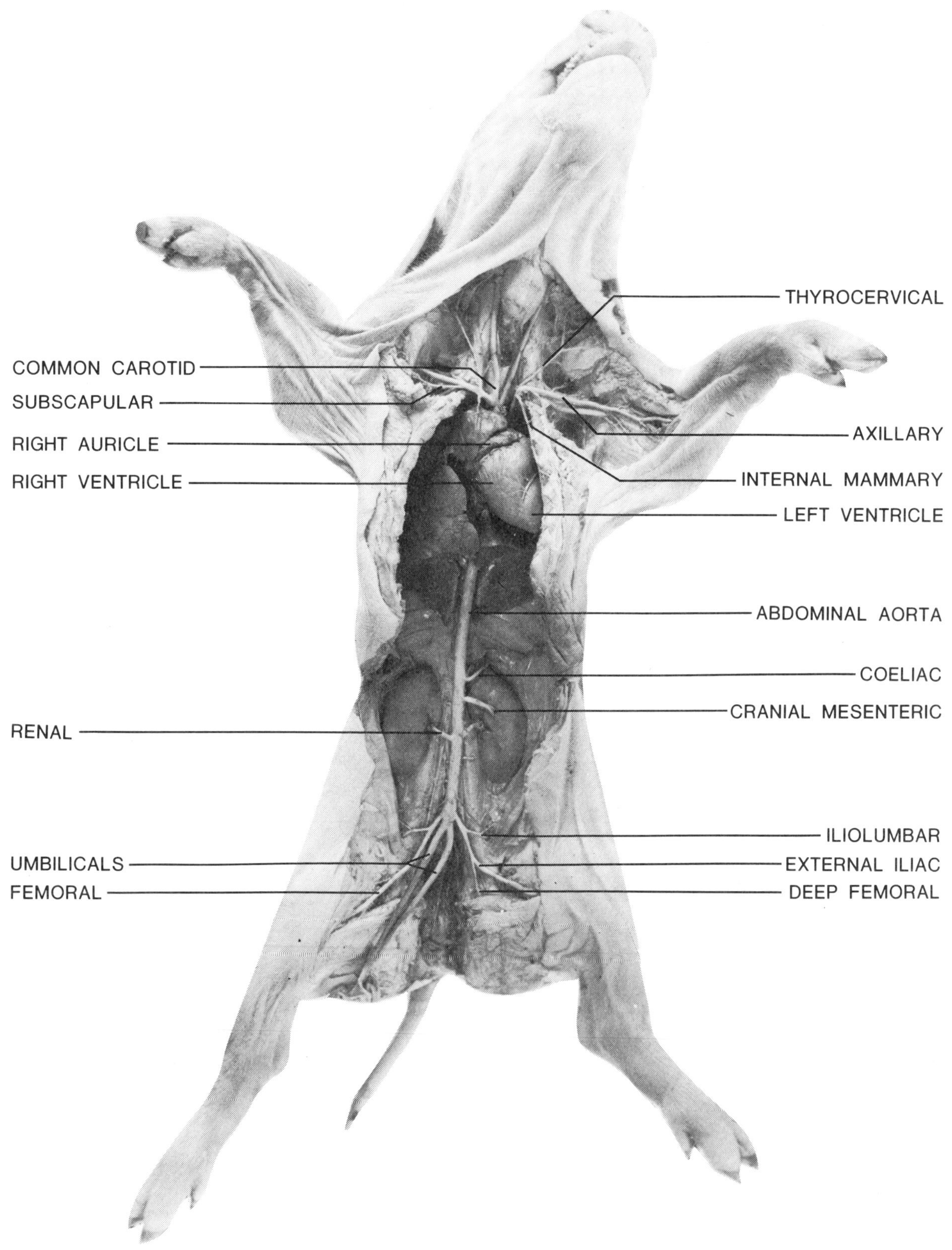

FIGURE 7.4. Arterial system, ventral view.

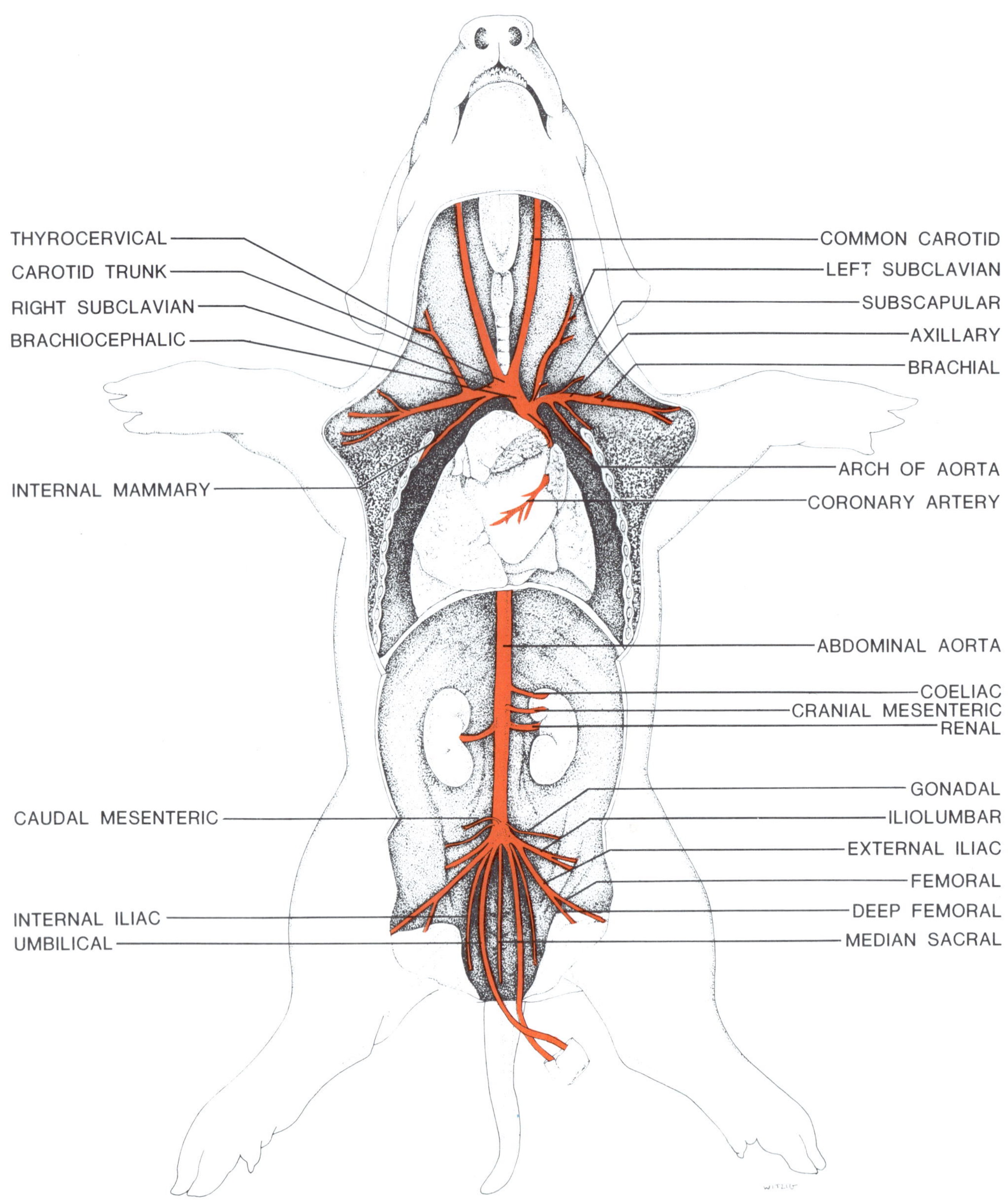

FIGURE 7.5a. Arterial system, ventral view.

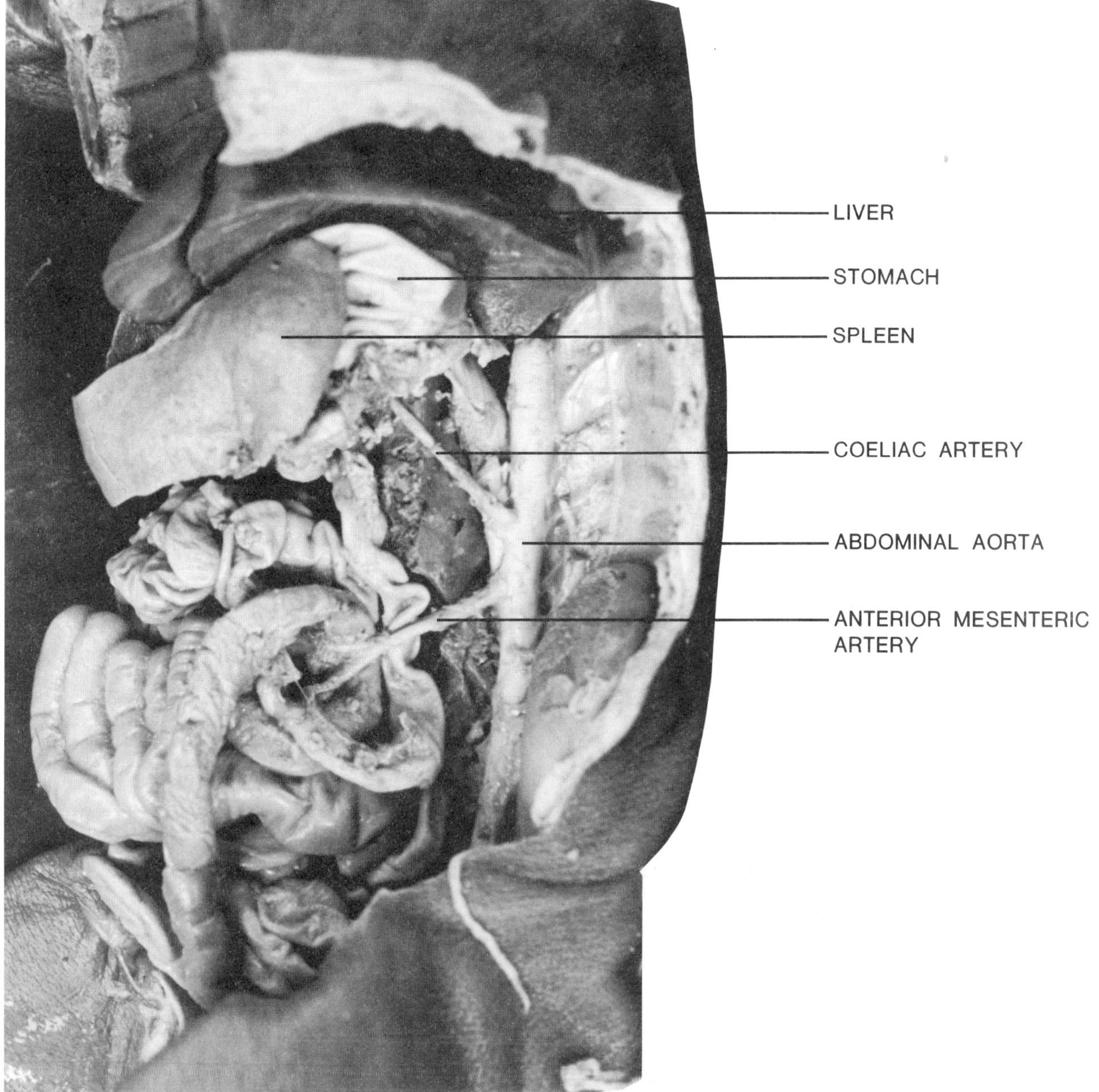

FIGURE 7.5b. Coeliac and anterior mesenteric arteries, ventral view.

7. *Right subclavian.* One of the two major branches of the brachiocephalic. Its branches are similar to those of the left subclavian.

8. *Costocervical trunk.* This branches from the dorsal surface of the subclavian at the level of the first rib and fairly close to the brachiocephalic. A major branch, the *vertebral* artery, proceeds from the costocervical trunk to pass anteriorly in the transverse foramina of the cervical vertebrae to the brain. Other branches of the costocervical need not be followed.

9. *Internal mammaries.* These originate from the subclavian at a point near the first rib and pass caudally in the thoracic wall near the midventral line. These vessels may have been cut when the thoracic cavity was opened.

10. *Subscapular.* A major branch of the axillary artery immediately external to the rib cage. This artery supplies the shoulder muscles.

11. *Thyrocervical.* Arises from the subclavian just cranial to the origin of the internal mammary to supply muscles of the chest and neck.

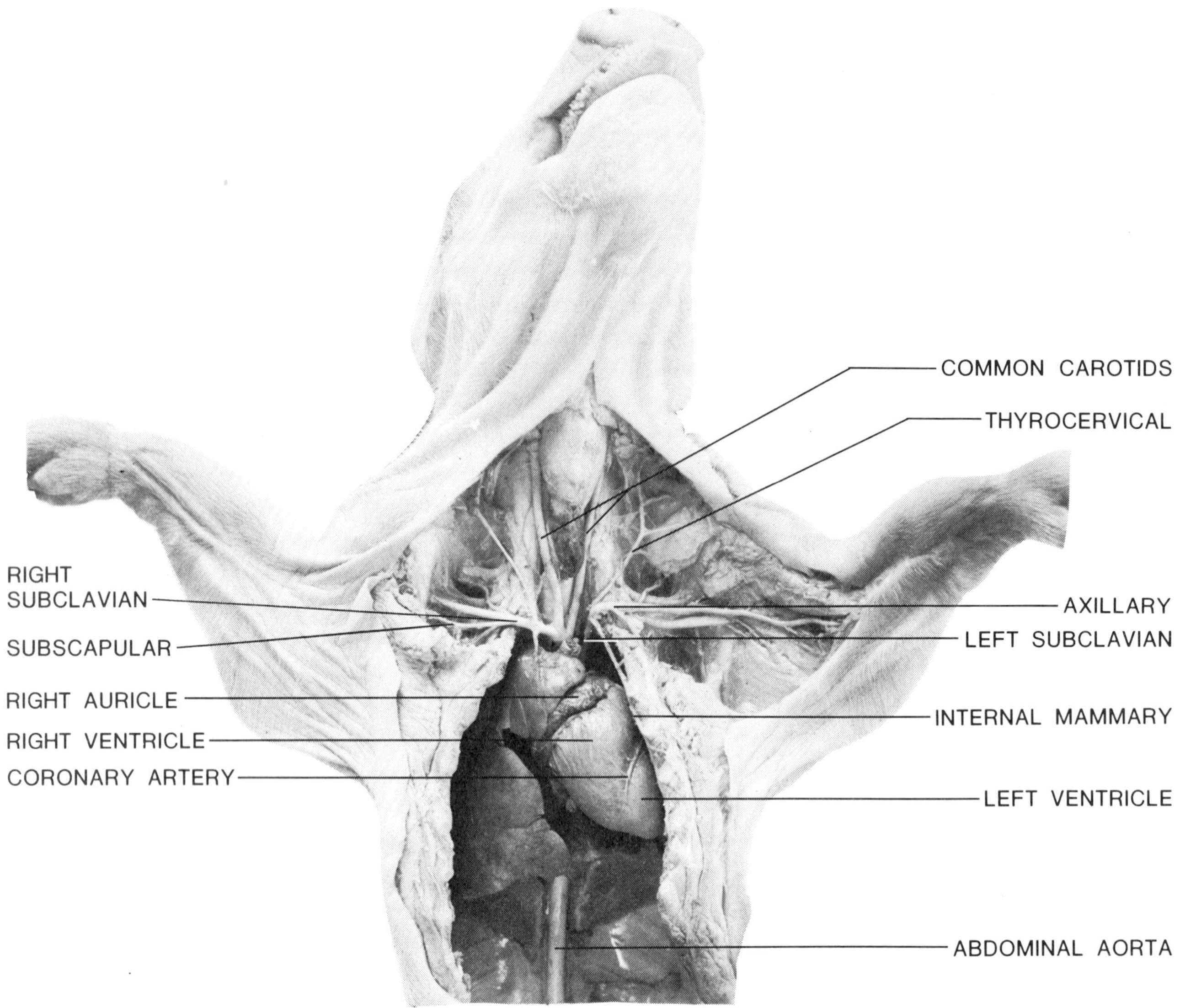

FIGURE 7.6. Cranial arteries, ventral view.

ARTERIES CAUDAL TO THE HEART
(Figures 7.1, 7.4, 7.5a, 7.5b, 7.7)

1. *Thoracic aorta.* The continuation of the aortic arch through the thoracic region. *Intercostal* arteries arise from the thoracic aorta to supply the intercostal muscles.
2. *Abdominal aorta.* The thoracic aorta after passing through the diaphragm into the abdominal cavity.
3. *Coeliac.* This is a fairly large unpaired branch arising from the aorta at a level with the cranial end of the kidneys. It will be necessary to remove muscle and connective tissue in this area before the vessel is clearly exposed. The coeliac artery divides into the *gastric* artery to the stomach, *splenic* artery to the spleen, and *gastrohepatic* artery to the stomach and liver.
4. *Cranial mesenteric.* An unpaired vessel arising from the aorta a short distance caudal to the coeliac artery. Branches of the cranial mesenteric supply the pancreas, small intestine, and large intestine.
5. *Renals.* These are paired vessels, right and left, to the kidneys.
6. *Lumbars.* Paired branches arising from the aorta at a level near the caudal end of the kidneys and supplying the muscles of the dorsal abdominal wall.
7. *Spermatics* (in the male). Paired vessels from the aorta proceeding through the inguinal canal together with the ductus deferens to the testes.
8. *Ovarian* (in the female). Paired vessels from the aorta to the ovaries.

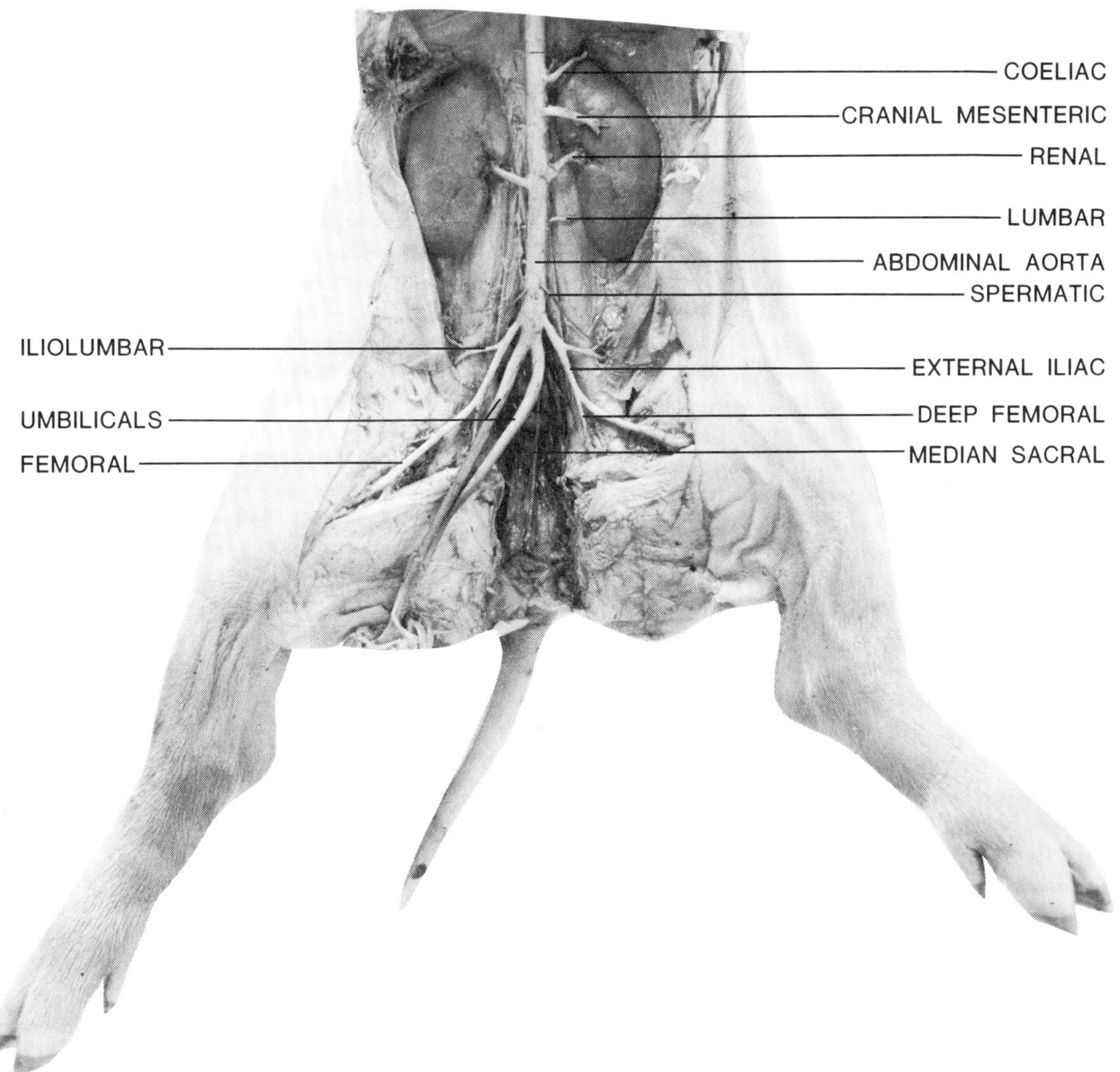

FIGURE 7.7.   Caudal arteries, ventral view.

9. *External iliacs.* Two large arteries proceeding into the hindlegs from the caudal end of the abdominal aorta. A pair of *iliolumbar* arteries branch from the iliacs to supply some of the muscles of the lower back. Each iliac continues into the leg where it gives rise to a small medial branch, the *deep femoral,* to the muscles of the thigh. Beyond the point of branching, the iliac is known as the *femoral,* dividing in the region of the knee into *popliteal* and *saphenous* arteries (not illustrated) to the lower leg.

10. *Caudal mesenteric.* A small unpaired vessel arising from the aorta between the iliacs and the gonadal arteries. It divides almost immediately with one branch proceeding cranially and one caudally to supply the large intestine.

11. *Internal iliacs.* The umbilical arteries appear to branch directly from the caudal end of the abdominal aorta. It is, however, the internal iliacs which branch from the aorta. They give rise to the umbilical arteries and then continue as much smaller vessels into the pelvic area.

12. *Median sacral.* A very small vessel from the caudal end of the aorta and continuing into the tail as the *caudal* artery.

EXTERNAL AND INTERNAL STRUCTURES OF THE HEART
(Figures 7.8, 7.9, 7.10, 7.11)

Free the heart from the body by cutting through the cranial and caudal venae cavae, the subclavians, common carotids, and the thoracic aorta just caudal to the heart.

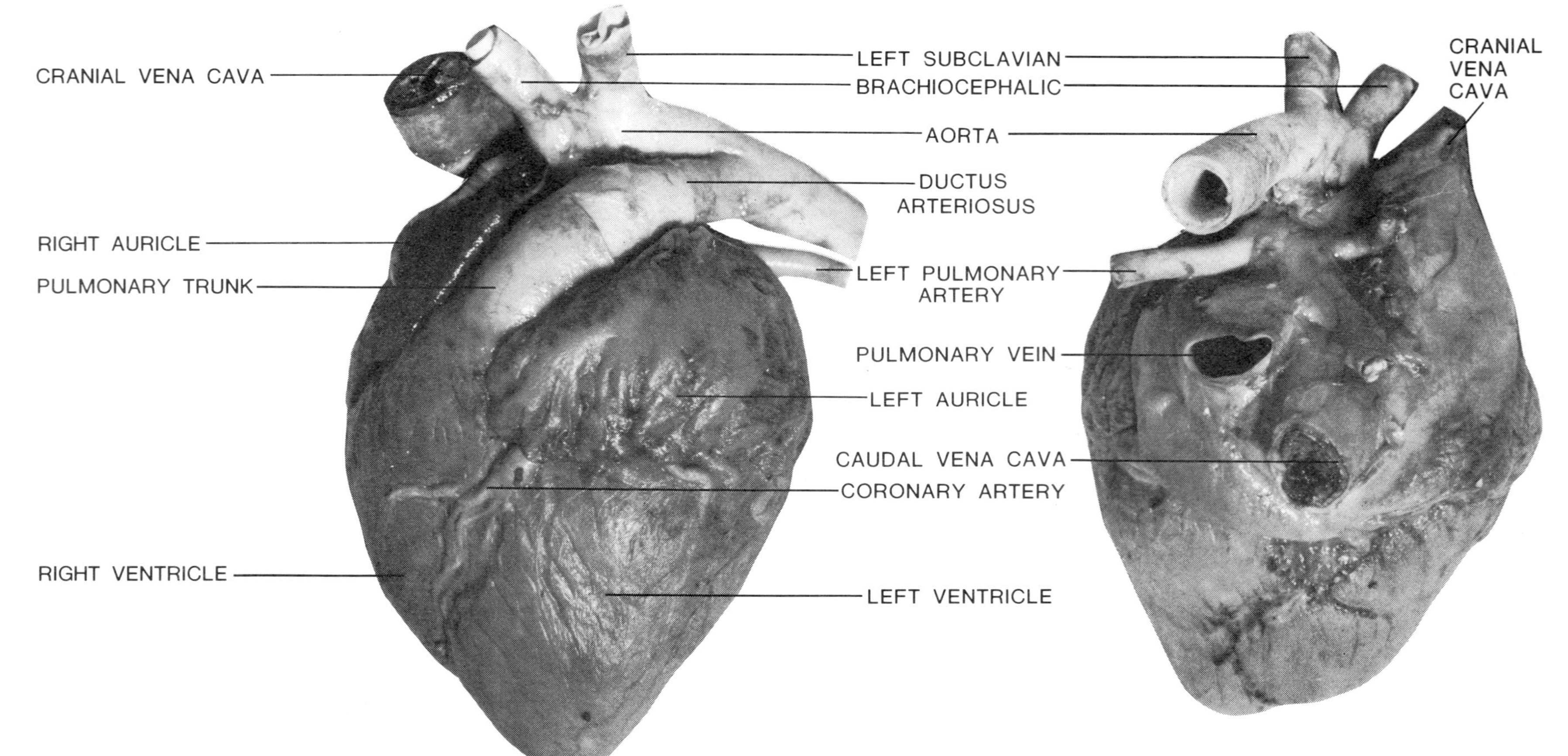

FIGURE 7.8. Fetal pig heart, ventral (left) and dorsal (right) views.

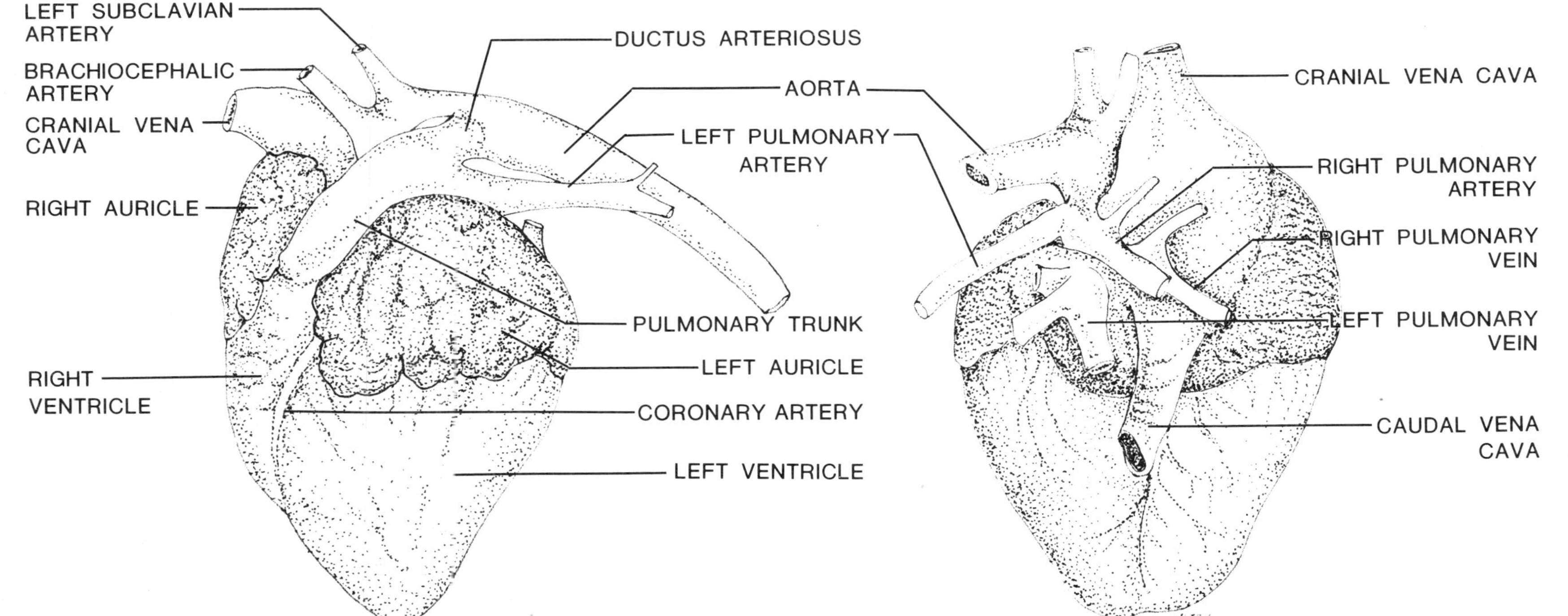

FIGURE 7.9.   Fetal pig heart, ventral (left) and dorsal (right) views.

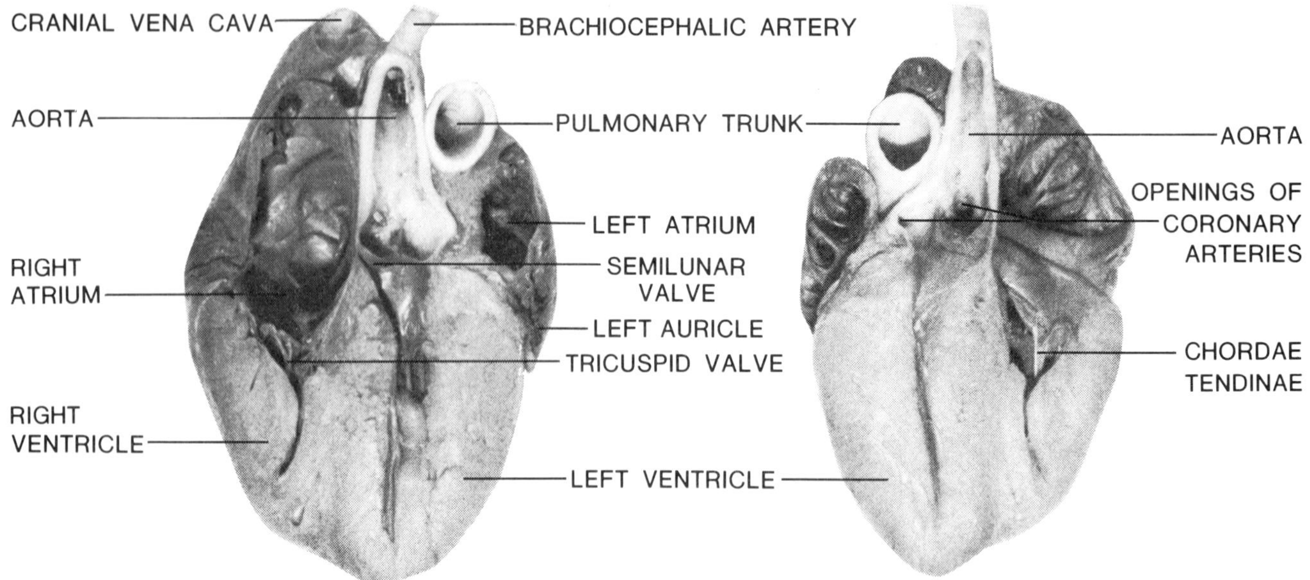

FIGURE 7.10.   Fetal pig heart, internal view.

Cut through the right and left pulmonary arteries at the point where they enter the lungs and through the pulmonary veins where they emerge from the lungs to the left atrium. Remove the heart from the thoracic cavity and proceed as follows.

With the ventral side of the heart uppermost, cut open the pulmonary trunk with a sharp razor blade and continue the cut down through the muscular wall of the right ventricle. Spread the vessel open and remove the latex. Identify three membranous cusps attached to the wall at the junction of artery and ventricle—these are the *semilunar valves*. The open ends of the valves face into the pulmonary trunk; thus, any backward flow of blood from the trunk into the ventricle will fill the valves with blood and prevent such backward flow.

Cut through the right auricle and atrium to expose the atrial cavity. Remove the latex and coagulated blood. Between the atrium and the right ventricular cavity are three cusps similar to the semilunar valves called *tricuspid* valves. The open ends of these face downward into the cavity of the ventricle. The internal walls of the ventricle have muscular ridges called *papillary* muscles and arising from them are fine fibers, the *chordae tendinae,* which are attached to the edges of the valves.

Cut open the left atrium and left ventricle in the same manner as you did for the other side of the heart. Between the left atrium and left ventricle are the *bicuspid* or *mitral* valves consisting of two cusps. Probe into the aorta from the exterior of the heart and note that the probe enters the cavity of the left ventricle. At the junction of the ventricle and aorta is a set of valves with three cusps—this is the *aortic semilunar.* Papillary muscles and tendinous cords are present as they were in the right ventricle. The wall between the two ventricles is the *septum.*

COMPARISON OF FETAL AND ADULT CIRCULATION
(Figure 7.12)

The fetus is dependent on the female during the period of intrauterine development (gestation) for food and oxygen and for the elimination of carbon dioxide and nitrogenous waste products. This dependence is reflected in the nature and functioning of certain parts of the circulatory system of the fetus as compared with circulation in the adult. The principal differences between the fetal and adult circulatory systems are seen in some of the blood vessels associated with the heart and lungs and with those vessels associated with the placenta and umbilical cord. Since the fetus has not breathed yet, the lungs are relatively compact structures and do not function in the exchange of respiratory gases. In the adult, nonoxygenated blood from the right atrium of the heart flows through the pulmonary trunk to the branching of that vessel to the lungs where the blood is oxygenated. In the fetus, some of the blood from the right atrium is directed into the right ventricle and out the pulmonary artery toward the lungs. Much of this blood is bypassed into the aorta through a shunt, the *ductus arteriosus,* and very little of it reaches the lungs.

Since the fetus has no provision for actively breathing on its own, it must depend on the female to supply oxygen and remove carbon dioxide. The place where the exchange of these gases takes place is the *placenta.* This is attached closely to the inner wall of the uterine horn and provides for a close *proximity* between fetal and maternal capillaries. Although there is a transfer of respiratory gases and food molecules and waste products of fetal metabo-

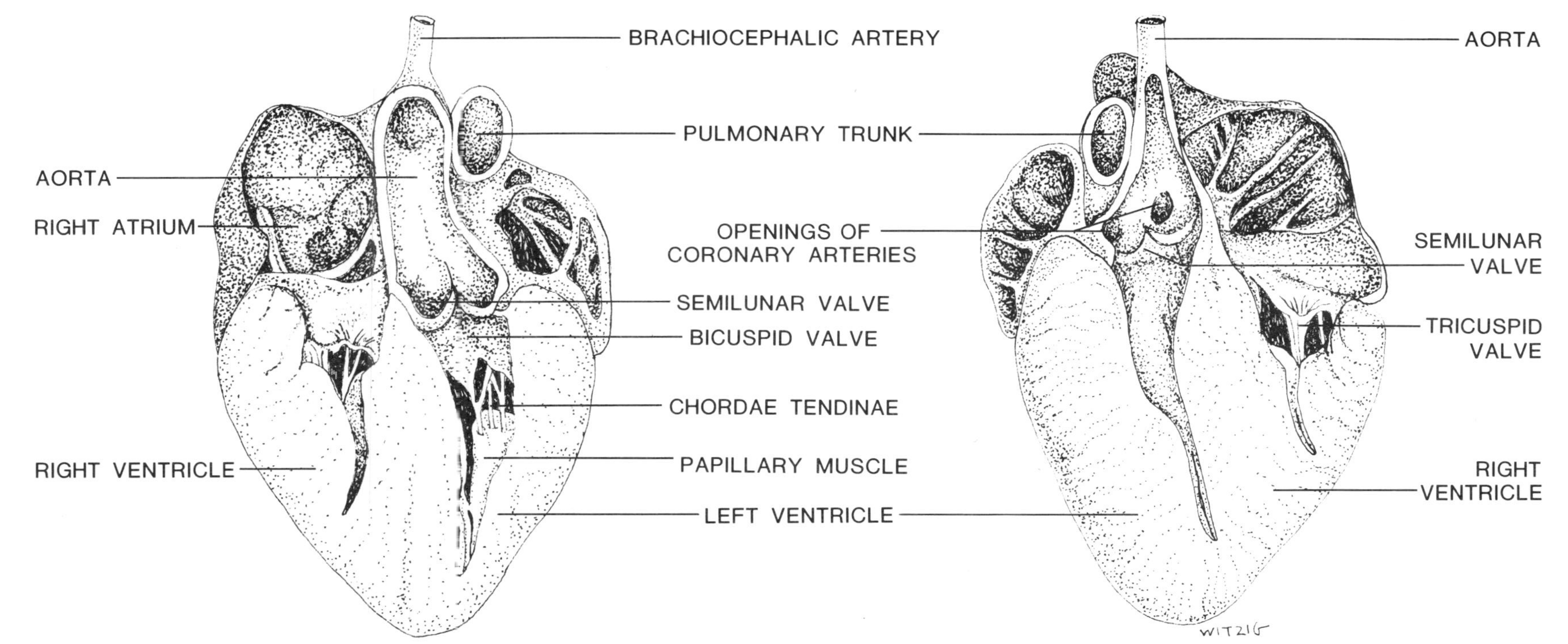

Fetal pig heart, internal view.

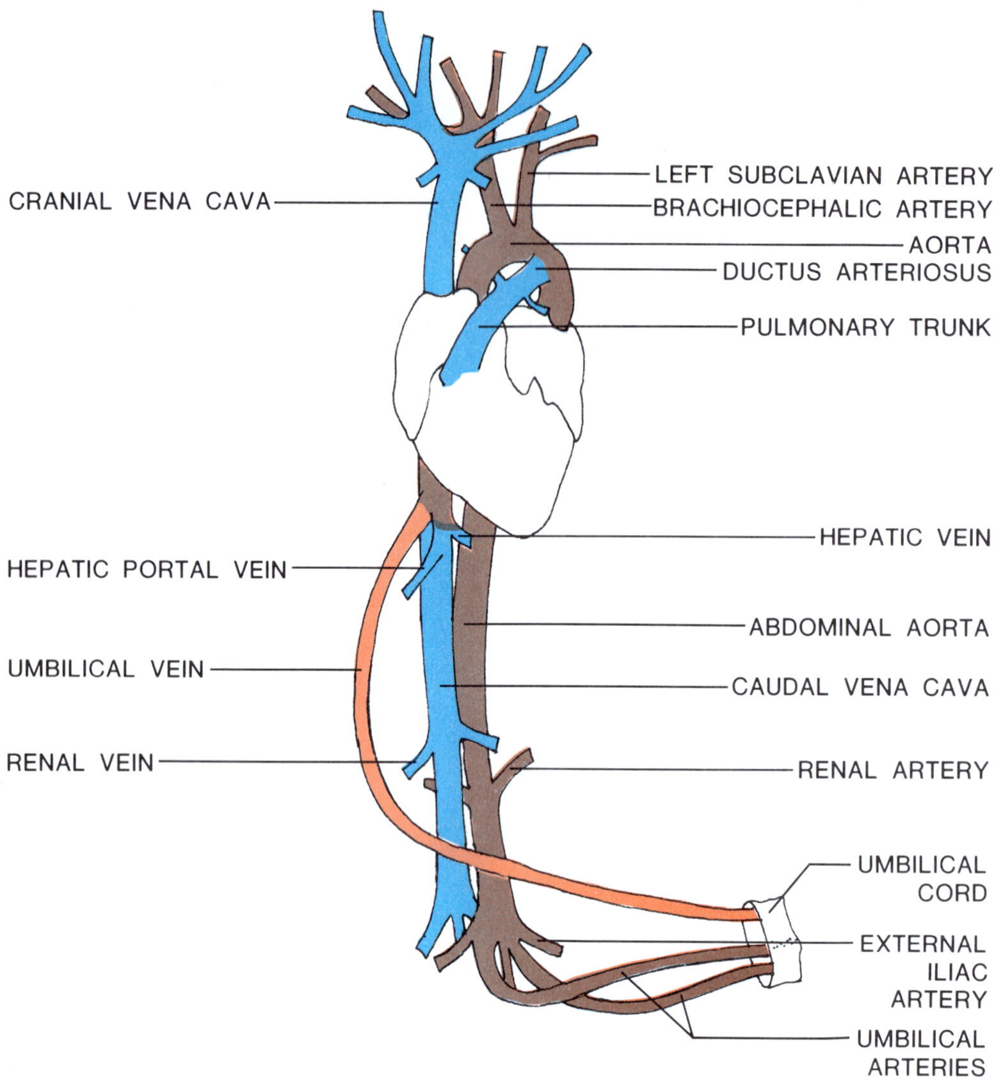

FIGURE 7.12.   Fetal circulation.

lism there is normally no mixing of the fetal and maternal blood within the placenta. Oxygen in the maternal bloodstream in the placenta is transferred to the umbilical vein of the fetus and is carried through a continuation of that vessel called the *ductus venosus* in the liver to finally empty into the caudal vena cava which enters the right atrium of the heart.

In a similar manner in which the umbilical vein carries oxygenated blood from the placenta to the fetus, the pair of umbilical arteries branching from the proximal portions of the internal iliacs carries nonoxygenated blood from the fetus to the placenta.

During fetal life, a temporary opening, the *foramen ovale,* exists between the right and left atria. This permits most of the oxygen-rich blood from the right atrium to pass into the left atrium and then to the left ventricle where it is pumped into the aorta to be distributed throughout the body of the fetus. During late fetal life, the ductus arteriosus becomes smaller in diameter resulting in some blood flow to the fetal lungs.

Following the birth of the fetus, and with the loss of the placenta, oxygenated blood no longer enters the right atrium. The umbilical vein becomes nonfunctional and that portion known as the ductus venosus through the liver is reduced to a solid cord, the *ligamentum venosum.* The umbilical arteries also become functionless. Within the heart, the foramen ovale closes, resulting in a complete separation between the two atrial chambers. Nonoxygenated blood from all parts of the body now returns to the right atrium via the cranial and caudal venae cavae and oxygenated blood from the lungs is directed into the left atrium through the pulmonary veins.

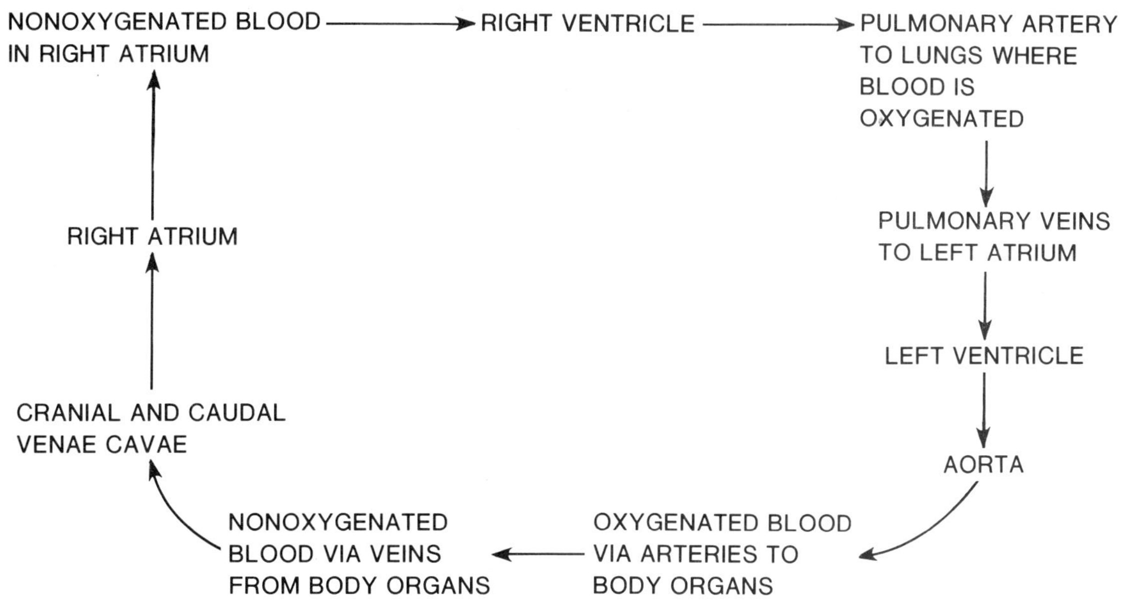

FLOWCHART OF ADULT CIRCULATION

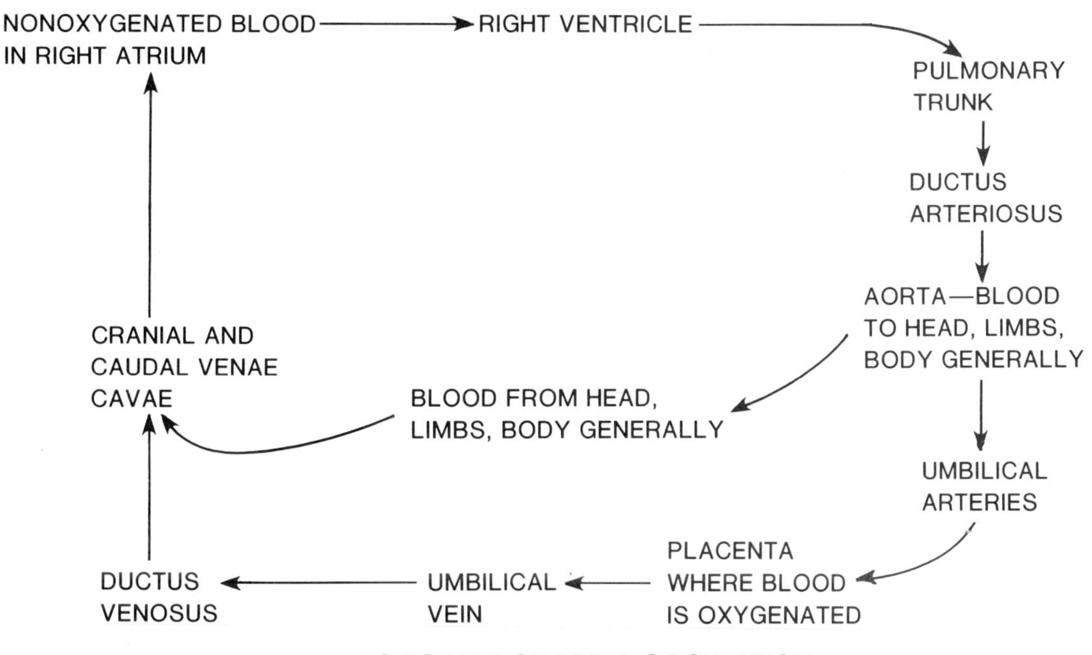

FLOWCHART OF FETAL CIRCULATION

The fetal bypass called the ductus arteriosus between the pulmonary trunk and aorta atrophies into a solid cord called the *ligamentum arteriosum* approximately eight weeks after birth.

Oxygenated blood from the lungs is now directed via pulmonary veins to the left atrium, to the left ventricle, and into the aorta for distribution to all parts of the body. Nonoxygenated blood from the body is eventually collected into the cranial and caudal venae cavae, then to the right atrium, to the right ventricle, and finally to the lungs via the pulmonary arteries for oxygenation.

Chapter 8
Respiratory System

(Figures 8.1, 8.2, 8.3, 8.4)

CHAPTER OBJECTIVES

1. To identify the gross features of the respiratory system.
2. To examine a microscopic section of lung tissue.

INTRODUCTION

The organs and structures that make up the respiratory system provide for the inspiration of air, the exchange of oxygen and carbon dioxide between the alveoli of the lungs and the blood capillaries, and the expiration of carbon dioxide.

1. *External nares* or nostrils.
2. *Paired nasal chambers.* Situated dorsal to the hard palate.
3. *Nasopharynx.* Situated dorsal to the soft palate.
4. *Glottis.* The opening from the pharynx into the larynx.
5. *Pharynx.* A common area into which the nasopharynx, oral pharynx, glottis, and esophagus open.
6. *Hyoid apparatus* (fig. 8.4). This is not part of the respiratory system but its proximity to the larynx makes this a convenient point to study the parts of the hyoid. Remove the muscles from the ventral throat region and expose a large cartilaginous structure, the *larynx.* Just cranial to it lies the hyoid apparatus consisting of a ventral *plate* or *body* with a pair of *cornua* or horns proceeding dorsally on either side. The shorter caudal horns lie on either side of the cranial end of the larynx, with the much longer cranial horns curving dorsally to terminate at the base of the tongue.
7. *Larynx.* Consists of four cartilages, the largest of which is the *thyroid* cartilage, a shield-shaped structure making up the ventral and lateral walls.

The *cricoid* cartilage lies just caudal to the thyroid cartilage, appearing somewhat as a ring with the broadest portion on the dorsal side. Free the larynx by cutting between it and the hyoid apparatus, and push the thyroid and cricoid cartilages to one side. Examine the dorsal surface of the larynx. Note the expanded portion of the cricoid. Clean away any muscle, mucous membrane, and connective tissue and identify a pair of smaller cartilages, the *arytenoids,* which project forward from the cranial edge of the cricoid to support the dorsal rim of the glottis. The fourth cartilage is the *epiglottis,* a white tablike structure attached to the ventral border of the glottis.

8. *Trachea.* Continues caudally from the larynx and divides into two major branches, the *bronchii,* to the lungs.* The trachea is supported throughout its length by a series of cartilaginous rings that are incomplete dorsally. The bronchii continue to divide into finer and finer branches, eventually ending as microscopic air spaces called *alveoli.* It is through the thin walls of the alveoli that the exchange of oxygen and carbon dioxide takes place. See figure 8.3.
9. *Lungs.* Cut through the trachea midway between the larynx and the lungs and remove it and the attached lungs from the body. Identify the *apical, cardiac, intermediate,* and *diaphragmatic* lobes of the right lung and the *cranial, cardiac,* and *diaphragmatic* lobes of the left lung.

    The space in the thoracic cavity where the lungs were removed is the *pleural* cavity, one on either side. These spaces are lined with *parietal pleura* and the lungs are covered with *visceral pleura.*

Figure 8.3 illustrates a microscopic section of the lung. The functional unit is the small thin-walled sac, the *alveolus,* whose walls are richly supplied with blood capillaries. It is through these thin walls that the exchange of oxygen and carbon dioxide takes place. A number of alveoli open into an *alveolar sac,* alveolar sacs into *bronchioles* which lead to *bronchii,* then to the trachea and eventually to the outside of the body through the internal and external *nares* or nostrils of the nose.

---

*Identify the esophagus, a soft muscular tube lying on the dorsal surface of the trachea.

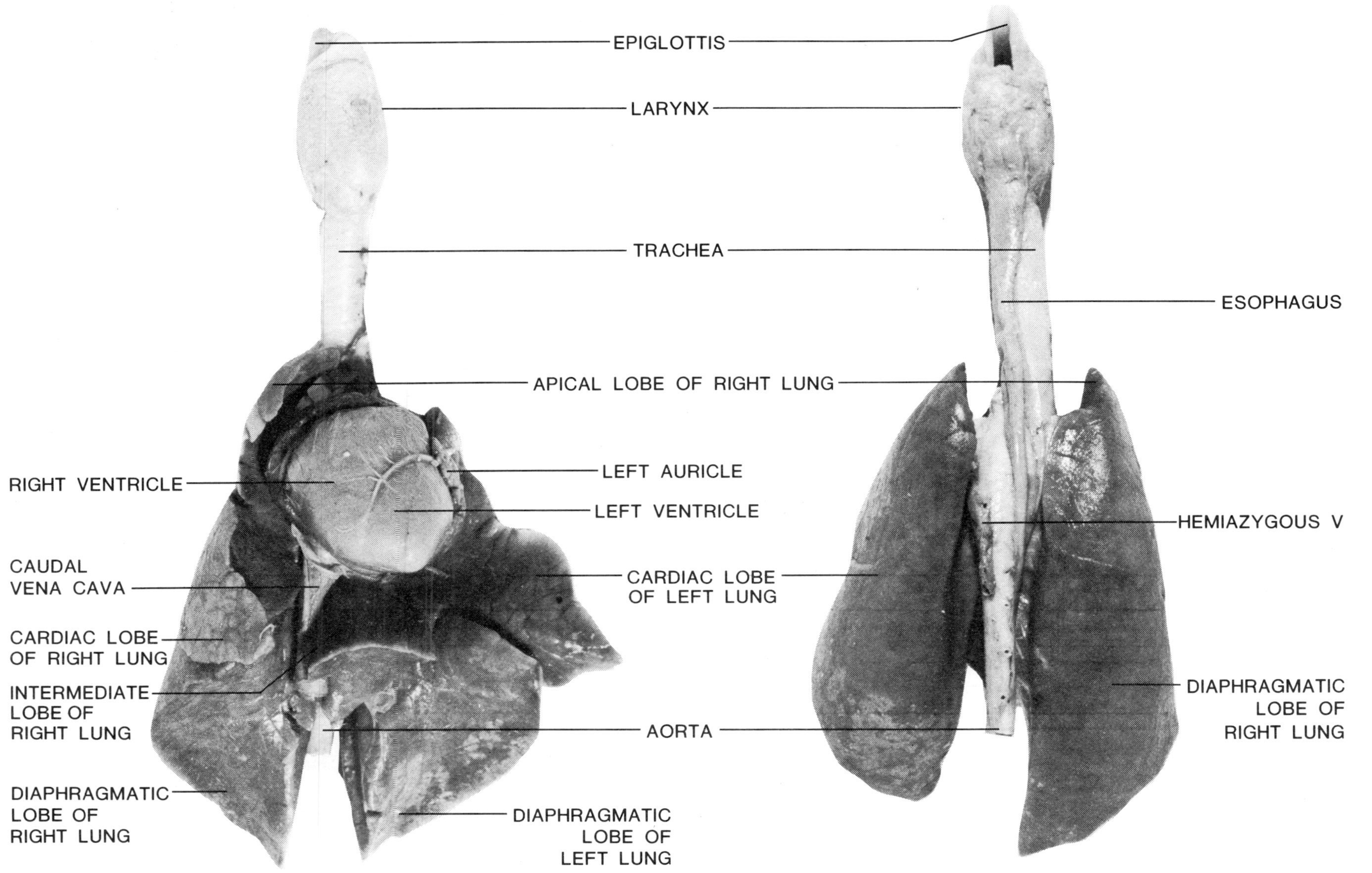

FIGURE 8.1. Respiratory system, ventral (left) and dorsal (right) views.

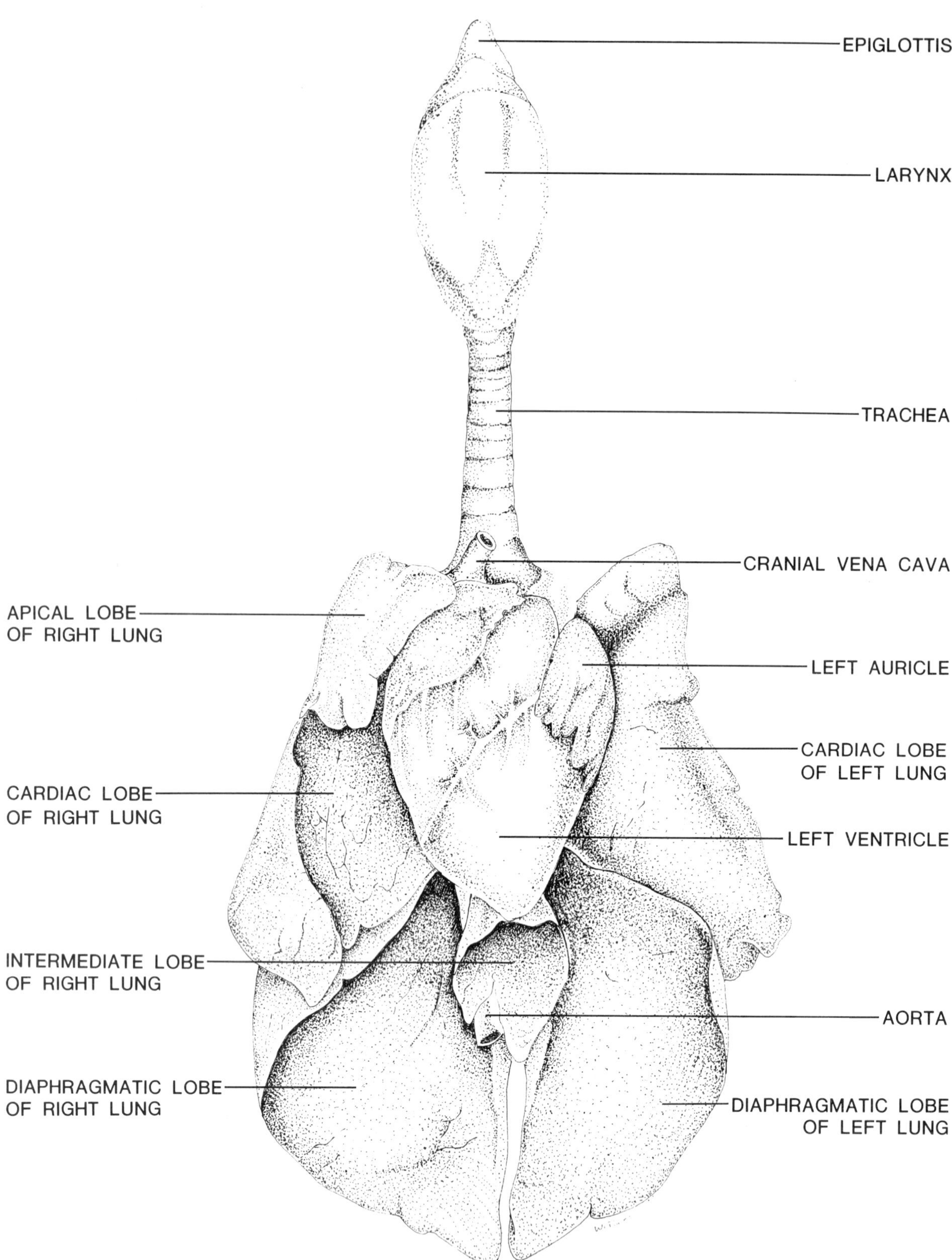

FIGURE 8.2. Respiratory system, ventral view.

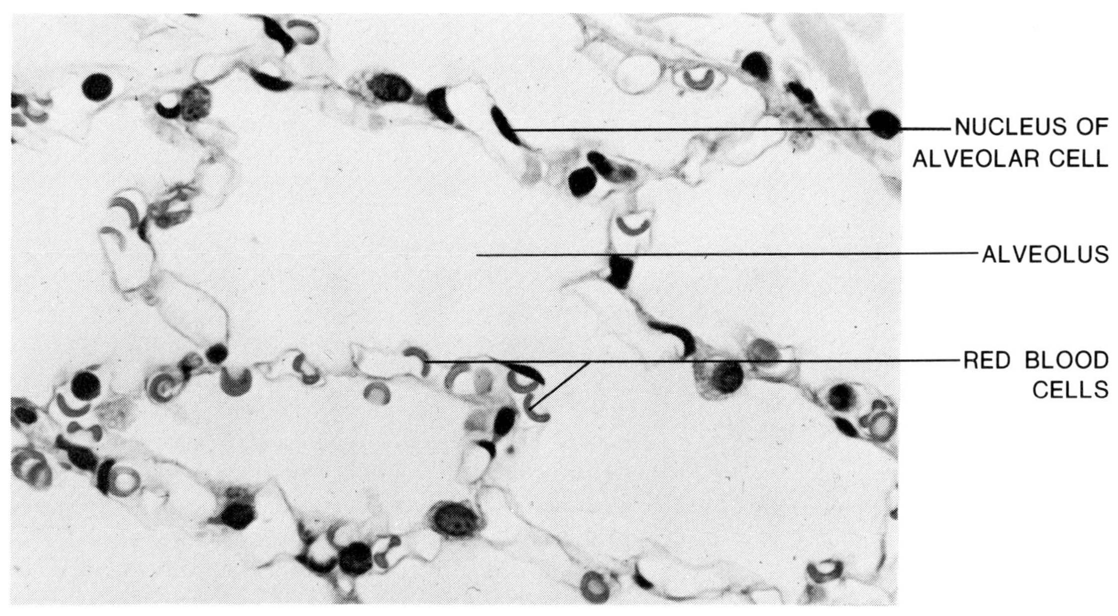

FIGURE 8.3. Photomicrograph of section of lung (×360).

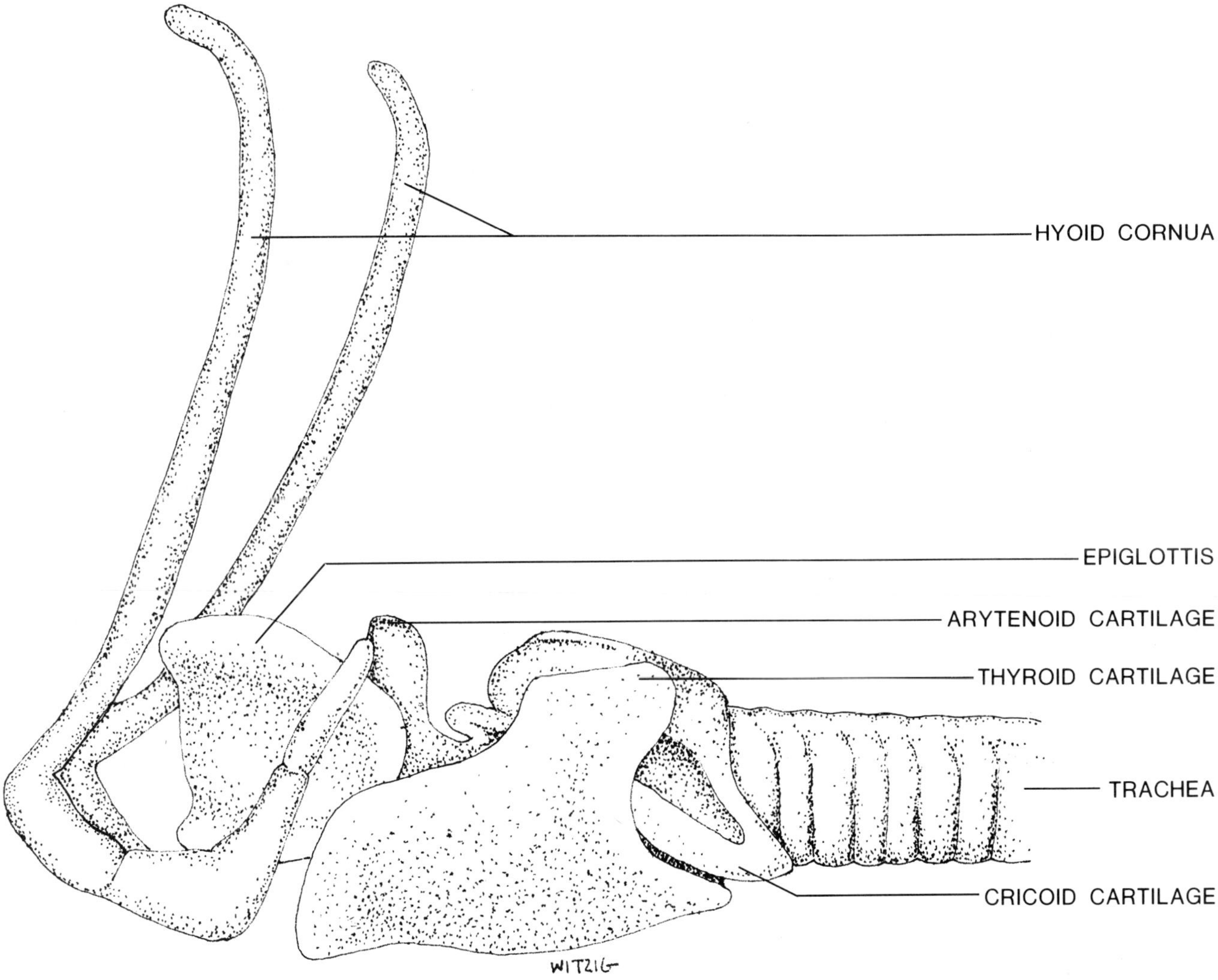

FIGURE 8.4. Laryngeal cartilages and hyoid bone.

Chapter 9
Nervous System

(Figures 9.1, 9.2, 9.3, 9.4, 9.5, 9.6, 9.7, 9.8, 9.9)

CHAPTER OBJECTIVES

1. To identify the major gross features of the brain and spinal cord.
2. To identify the cranial and spinal nerves.
3. To identify the sympathetic portion of the autonomic nervous system.

INTRODUCTION

The nervous system adapts the organism to its external environment through organs of special sense such as the eye and ear (*exteroceptors,* chapter 10) and to its internal environment through *interoceptors* in the viscera.

The central nervous system is composed of the brain and spinal cord; the peripheral nervous system consists of cranial and spinal nerves and their branches; the autonomic system has two major divisions: the sympathetic and parasympathetic portions, which control smooth muscles, glands, and viscera. (The sheep brain may be substituted for a dissection of the fetal pig brain. If the sheep brain is used, follow the directions on page 99.)

SPINAL CORD AND SPINAL NERVES
(Figures 9.1, 9.2, 9.3, 9.4)

Remove the muscles that lie dorsal to the vertebral column as well as those muscles lying a short distance to either side of the column. With a sharp scalpel, gradually shave down the spines and neural arches of the vertebrae until the spinal cord is exposed. Three *meninges* or coverings surround the cord; the most readily apparent is the outermost layer, the *dura mater.* This is a tough, transparent connective tissue sheath that must be slit in order to expose the spinal cord. The middle covering is the *arachnoid* layer, which will not be seen. The innermost meninx is the *pia mater* closely applied to the substance of the spinal cord. At the level of the forelegs and hindlegs, the spinal cord is enlarged. These are the *cervical* and *lumbar enlargements,* respectively, and result from an increase in the number of nerve cells in these two regions supplying innervation to the appendages. At its cranial end, the cord gradually widens to become the most caudal part of the brain, the *medulla oblongata.* At its caudal end, the spinal cord narrows to a thin strand or filament called the *filum terminale.*

There are 33 pairs of spinal nerves associated with the spinal cord. Eight of these nerves are cervical, 14 thoracic, 7 lumbar, and 4 sacral. Each nerve is formed by the union of a *dorsal root* and a *ventral root.* Each root arises as a series of small roots or fibers from the *caudal lateral sulcus* on each side of the dorsal surface of the cord. The dorsal root, conveying sensory impulses into the spinal cord, can be identified by the presence of a distinct swelling called the *dorsal root ganglion.* The ventral root, carrying motor impulses from the cord to some effector, such as a muscle, has no external ganglion and is not as readily observed as is the dorsal root. Just caudal to the lumbar enlargement, the spinal nerves do not proceed laterally but turn caudally on either side of the filum terminale. This group of nerve fibers is known as the *cauda equina* because of their resemblance to a horse's tail.

It is tedious and time-consuming for each student to expose the full length of the spinal cord with the associated nerves. This dissection might be shared by having one group of students dissect one region while other groups dissect other regions of the spinal cord.

In two regions, the spinal nerves form interconnecting networks of fibers called *plexi.* One is the *brachial* plexus supplying the forelimb, the other is the *lumbosacral* plexus supplying the hindlimb. Many of the nerves comprising the brachial plexus were seen previously as white cords accompanying the arteries and veins into the foreleg. To expose the lumbosacral plexus, slit the peritoneum on the dorsal body wall between the caudal end of the kidney and the *groin.* In the groin, lateral to the external iliac artery, find 3 to 4 small white fibers that proceed into the muscles on the medial surface of the thigh. Follow the branches cranially and note that they unite into a larger trunk that emerges from the spinal cord in the lumbar region. Do not confuse this nerve trunk with a long, thin, shiny tendon that lies between the nerve and the external iliac artery.

The *autonomic* nervous system consists of a complex of nerve fibers and ganglia with connections to the brain and spinal cord. It consists of the *sympathetic* and *parasympathetic* portions, both of which are involved in the control of smooth muscles, glands, and visceral organs. The sympathetic trunk will be seen by looking on either side of the middorsal line immediately adjacent to the thoracic and abdominal aorta. It will appear as a thin fiber interrupted at intervals by a series of small swellings, the ganglia.

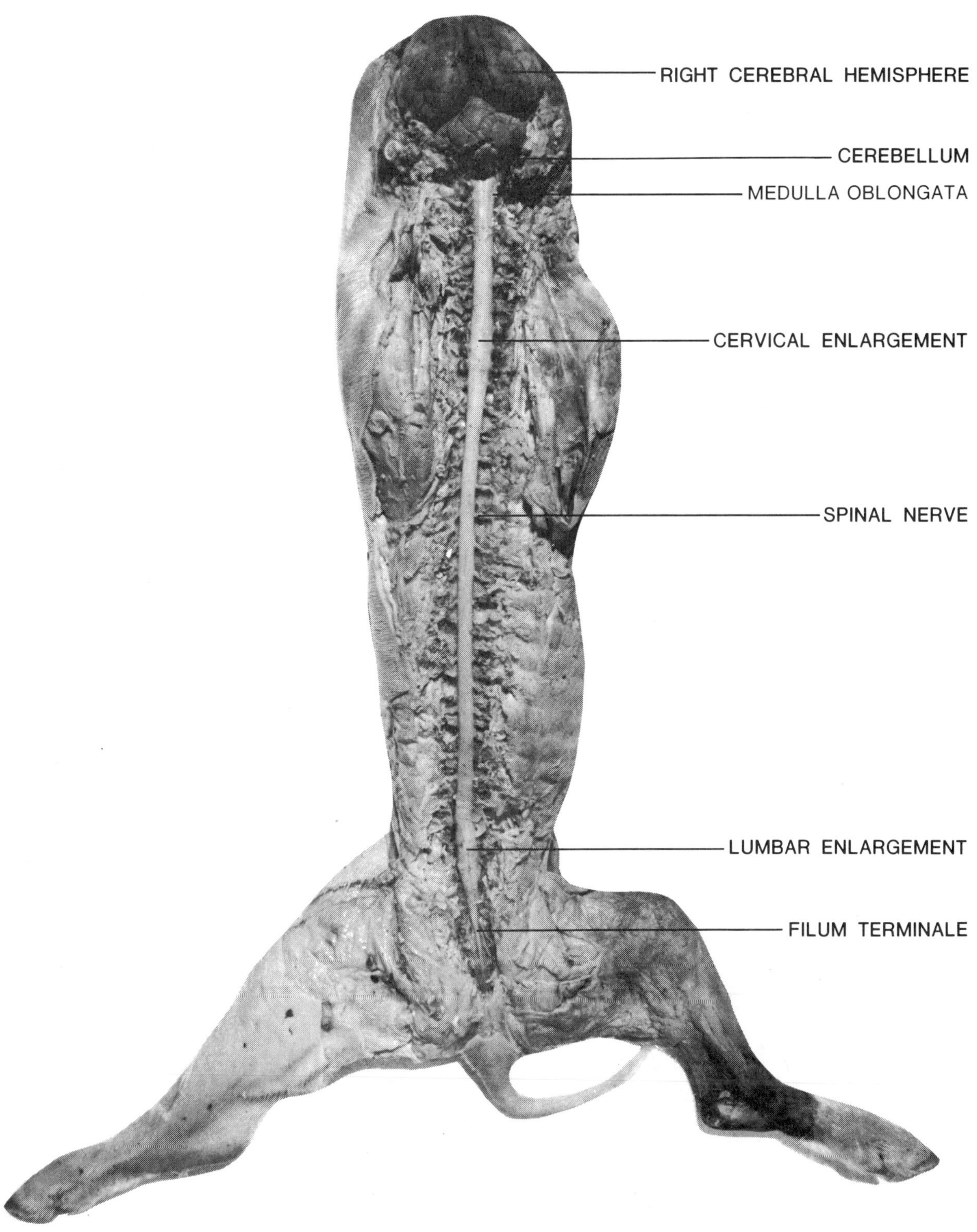

FIGURE 9.1. Brain, spinal cord, and spinal nerves, dorsal view.

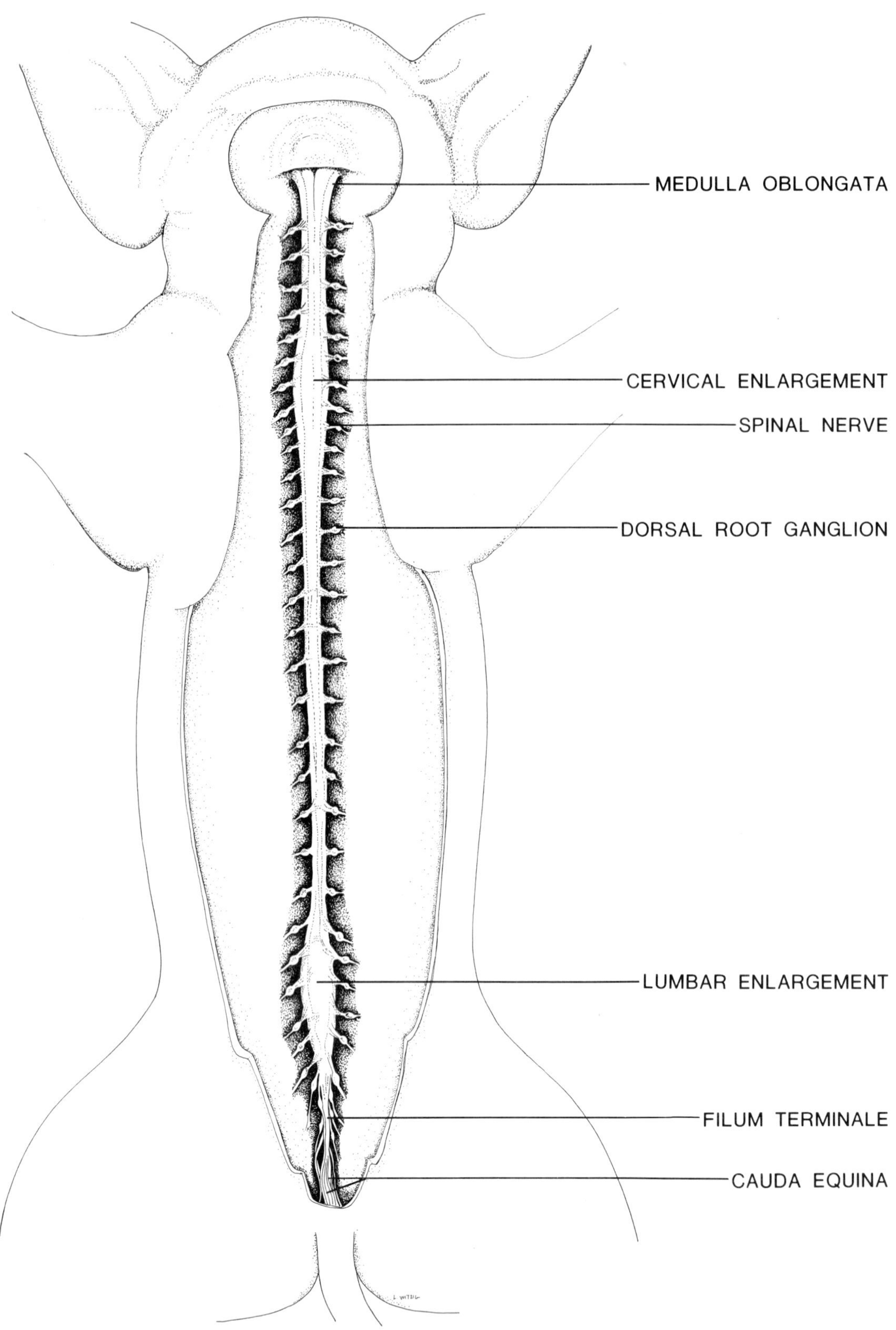

FIGURE 9.2.   Spinal cord and spinal nerves, dorsal view.

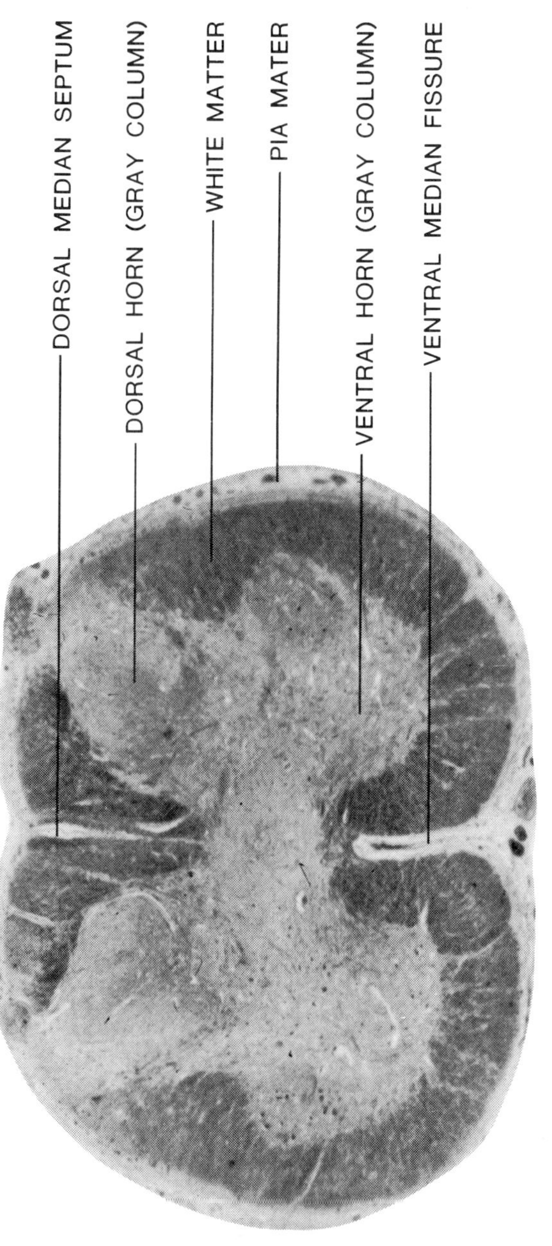

FIGURE 9.3. Cross section of spinal cord (photomicrograph ×10).

BRAIN
(Figures 9.1, 9.5, 9.6, 9.7)

Both sheep brain and fetal pig brain can be used for a study of the structures of this part of the central nervous system. The formalin-hardened sheep brain will probably prove to be more satisfactory for identifying the cranial nerves and hypophysis since these structures are usually lost in removing the fetal pig brain from the cranial cavity. All other parts of the brain can be studied using the fetal pig brain.

Make a longitudinal cut through the skin and muscles of the head on the dorsal side from the base of the snout to the base of the skull. From the longitudinal incision, make a transverse cut ventrally to the angle of the jaws in front and similar cuts at the caudal end of the longi-

tudinal incision to a level just ventral to the ears. Remove the skin and muscle to expose the skull.

The brain can now be exposed by removing the skull in small sections. Make a longitudinal cut along the mid-dorsal line of the skull, taking care not to cut too deeply. Make two lateral cuts about one inch apart and connect these with the longitudinal cut. Break off the small pieces of skull and continue in the same manner until the entire dorsal and lateral areas of the skull have been removed and the brain is revealed. In the caudal part of the skull, it will be necessary to dissect away a large mass of muscle as well as the heavy *occipital* bone through which the spinal cord passes. The tough membrane immediately under the skull and enclosing the brain is the *dura mater.* This should be removed to expose a thinner and more fragile membrane, the *pia mater* on the surface of the brain. As with the spinal cord, the middle meninx, the *arachnoid* layer, will not be seen.

The gross features of the brain can now be identified. They are the large right and left *cerebral hemispheres,* separated by a *longitudinal groove;* a smaller mass, the *cerebellum,* caudal to the cerebrum; and the *medulla oblongata,* caudal and ventral to the cerebellum and continuous with the spinal cord.

The convoluted surface of the cerebrum shows ridges called *gyri* and depressions called *sulci.* These make up the *cortex* or gray matter composed of nerve cells and supporting cells called *neuroglia.* Underlying the cortex is the white matter consisting of nerve fibers enclosed in a fatty material known as *myelin.* Recall the relative position of the white and gray matter in the spinal cord where the white matter was on the outside and the gray matter within.

Spread apart the cerebral hemispheres and look deeply between them. The white mass that will be seen is the *corpus callosum,* a bundle of fibers connecting the two hemispheres. The cerebrum is separated from the cerebellum by a deep cleft, the *transverse fissure.* The cerebellum has three parts, the central or median *vermis* and two lateral *hemispheres.* The most caudal part of the brain is the medulla oblongata. Lift the caudal edge of the cerebellum and observe that the medulla is roofed over with a thin vascular membrane, the *posterior choroid plexus.* Upon removing the plexus, the space that is revealed is the *fourth ventricle.*

Spread apart the cerebral hemispheres from each other and from the cerebellum. The four rounded bodies that are exposed constitute the *corpora quadrigemina.* The cranial pair are the *superior colliculi,* the caudal pair the *inferior colliculi.* Lying in the midline just cranial to the superior colliculi is a small structure, the *pineal body.*

Turn to the ventral side of the brain and identify the following structures.

A pair of *olfactory bulbs* on the cranioventral part of the cerebrum. A band of fibers, the *olfactory tracts,* pro-

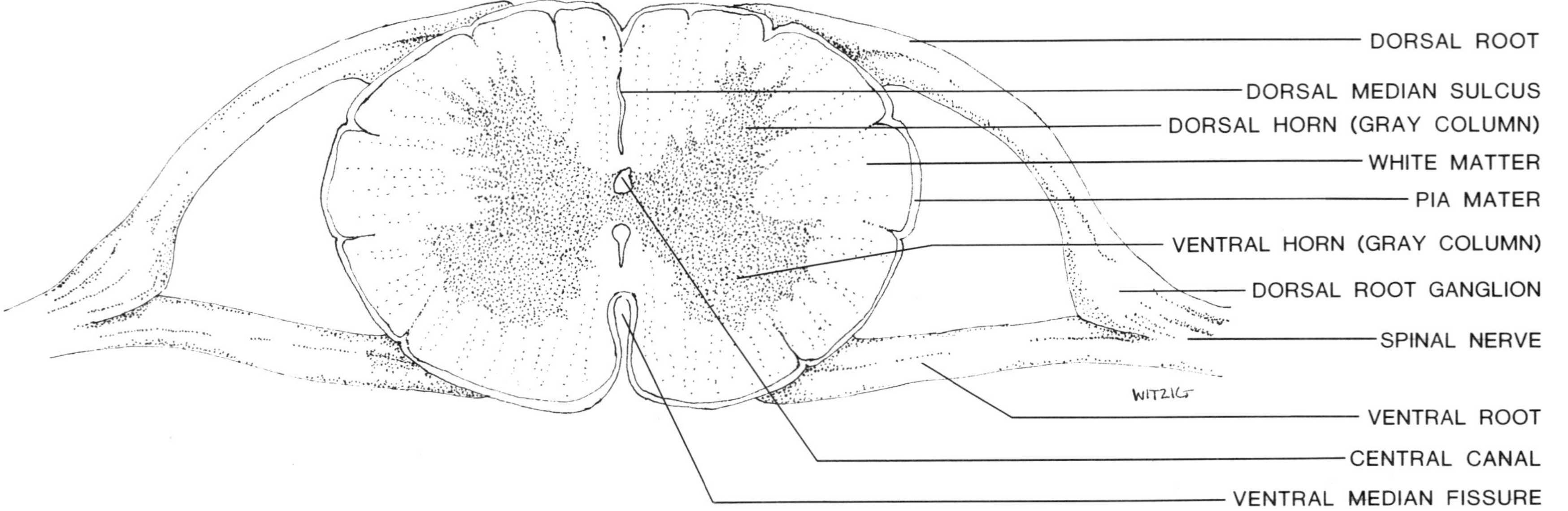

FIGURE 9.4. **Cross section of spinal cord (diagrammatic).**

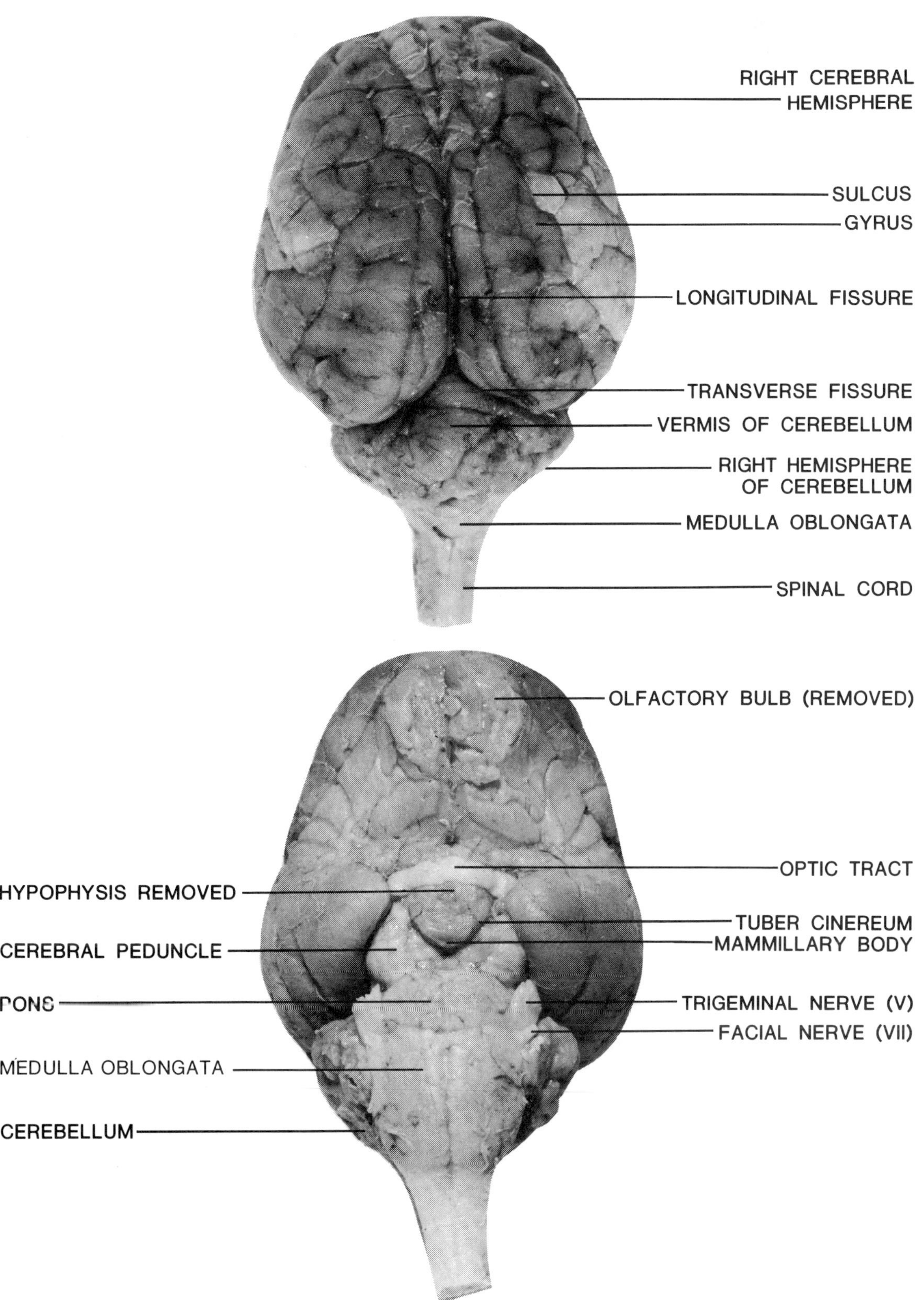

FIGURE 9.5. Fetal pig brain, dorsal (above) and ventral (below) views.

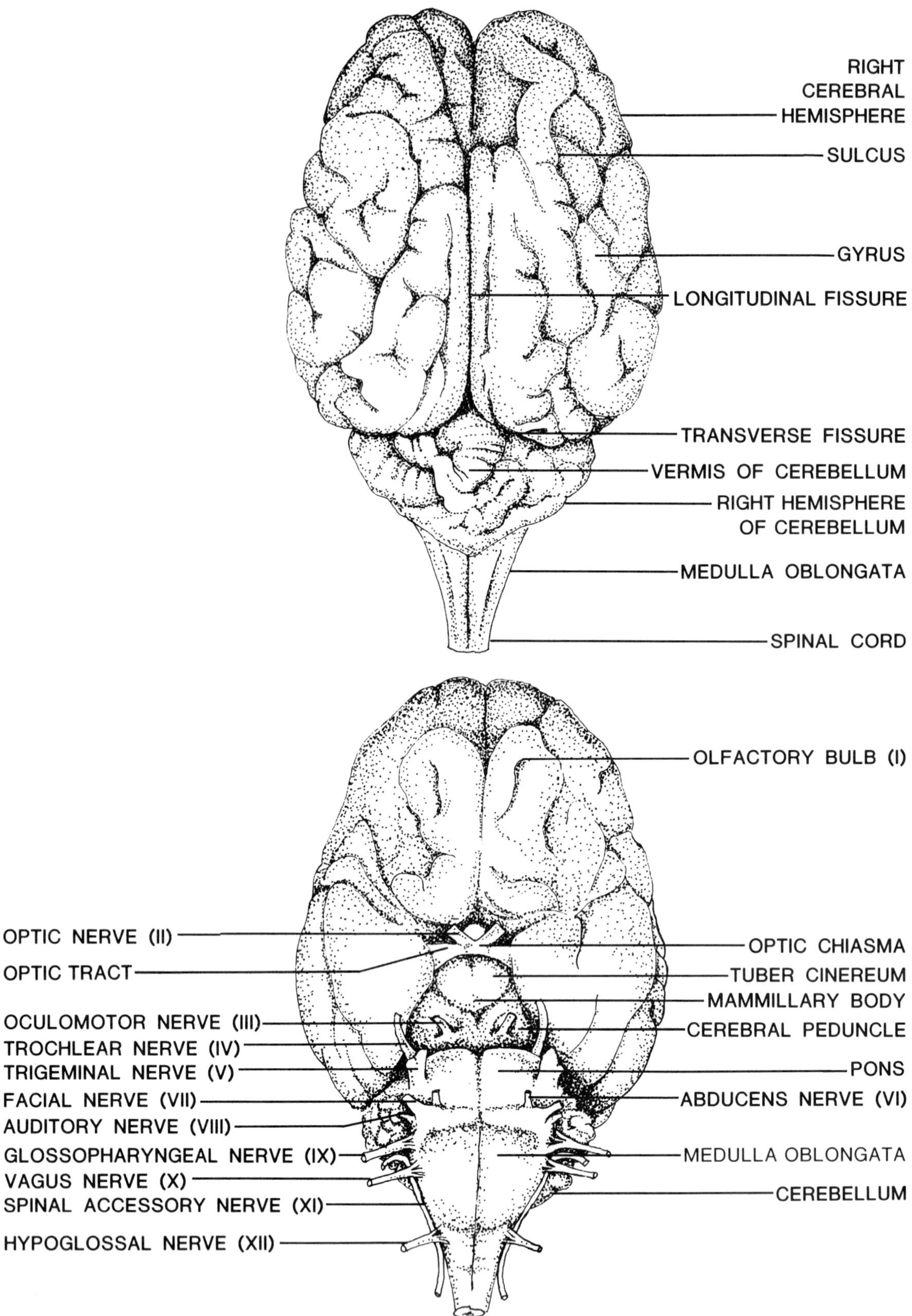

FIGURE 9.6. Fetal pig brain, dorsal (above) and ventral (below) views.

NUMBER AND NAME	ORIGIN ON BRAIN	ACTION	DISTRIBUTION
I Olfactory	Olfactory bulb	Sensory	Olfactory cells
II Optic	Thalamus	Sensory	Retina of eye
III Oculomotor	Cerebral peduncle	Motor	Muscles of eyeball (superior, inferior, and medial recti and inferior oblique)
IV Trochlear	Anterior and lateral to pons	Motor	Muscle of eyeball (Superior oblique)
V Trigeminal	Posterior and lateral border of pons	Mixed	Skin of head, muscles of jaw and tongue
VI Abducens	Midventral posterior edge of pons	Motor	Muscle of eyeball (Lateral rectus)
VII Facial	Posterior to trigeminal	Mixed	Muscles of expression and mastication
VIII Auditory	Posterior to facial	Sensory	Organ of Corti of inner ear
IX Glossopharyngeal	Near vagus	Mixed	Tongue and pharynx
X Vagus	Lateral border of medulla oblongata near VIII	Mixed	Heart, lungs, larynx, stomach, intestine
XI Spinal Accessory	Lateral border of medulla oblongata and anterior end of cord	Mixed	Neck and shoulder muscles
XII Hypoglossal	Medulla posterior to X	Mixed	Muscles of tongue

FIGURE 9.7. Cranial nerves.

ceeds caudally from each bulb. Two large trunks, the *optic tracts,* cross in the midline of the cerebrum; the crossing is the *optic chiasma.* Just caudal to the chiasma is a slit marking the point of attachment of the pituitary and a median swelling, the *mammillary* body. Two larger swellings lateral and caudal to the latter are the *cerebral peduncles.* These are motor tracts that convey nerve impulses from the cortex of the brain to the origin of various nerves in the spinal cord and brain. Next caudally is a wide transverse band of fibers, the *pons,* which connects the two hemispheres of the cerebellum. The narrow transverse band just caudal to the pons is the *trapezoid* body, which is associated with hearing. The remainder of the brain to its junction with the spinal cord is the medulla oblongata.

SAGITTAL SECTION OF BRAIN
(Figures 9.8, 9.9)

Make a midsagittal section through the entire brain and identify the following.

1. *Cerebellum.* The branching arrangement is the *arbor vitae* composed of myelinated nerve fibers.
2. *Medulla oblongata.* Ventral to the cerebellum and showing the cavity of the fourth ventricle.
3. *Superior* and *inferior colliculi* of the corpora quadrigemina.
4. *Pineal body* or *epiphysis* lying just cranial to the superior colliculi.
5. *Corpus callosum.* An elongate white band at the bottom of the longitudinal fissure. It is composed of the *rostrum* directed cranially; the *genu* caudal to the rostrum; the *body;* the caudal *splenium.* Extending ventrally from the body is a band, the *fornix.* The *lateral ventricles,* one in each of the cerebral hemispheres, lie on either side of a thin membrane, the *septum pellucidum,* which extends between the rostrum and the fornix. The lateral ventricles are connected to the *third* ventricle by the *Foramina of Monro.* The third ventricle, in turn, is connected to the fourth ventricle by the *cerebral aqueduct* or *Aqueduct of Sylvius.*
6. *Mass intermedia.* A round mass just cranial to the pineal body and constituting the *middle commissure* of the *thalamus.*
7. *Optic chiasma.*
8. *Sulci* and *gyri* of the cerebral hemispheres.

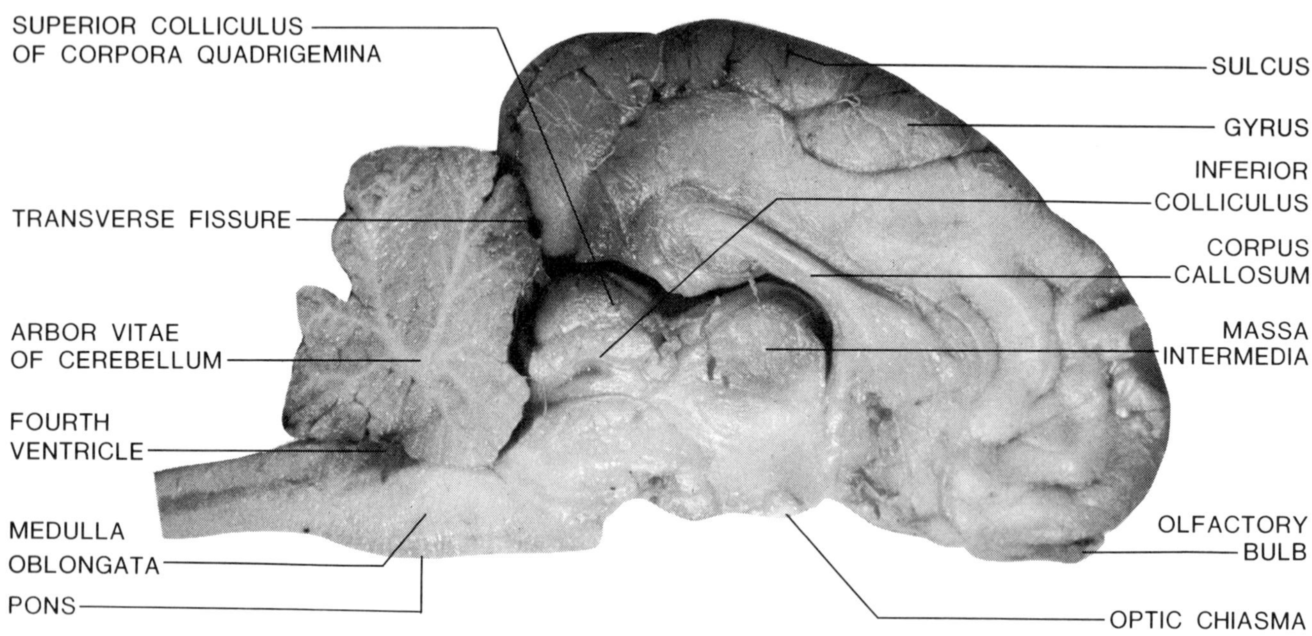

FIGURE 9.8. Fetal pig brain, sagittal section.

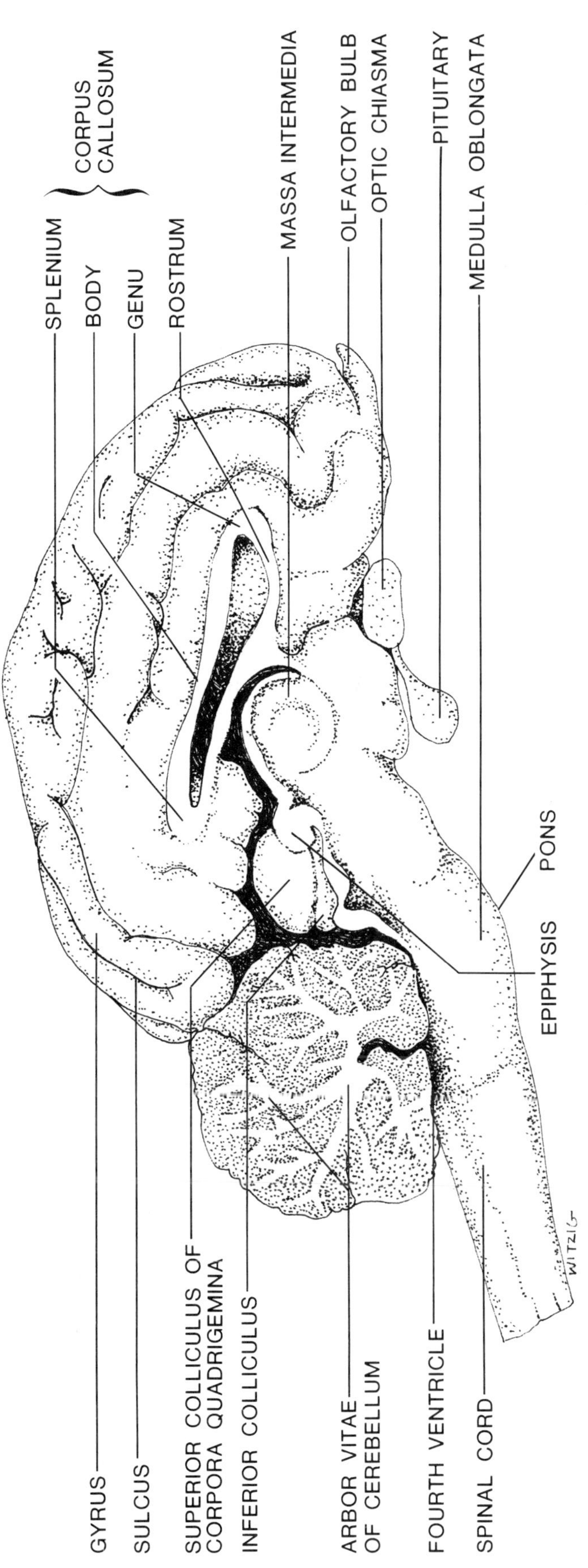

FIGURE 9.9. Fetal pig brain, sagittal section.

Chapter 10
Sense Organs

(Figures 10.1, 10.2, 10.3, 10.4, 10.5, 10.6, 10.7, 10.8)

CHAPTER OBJECTIVES

1. To identify the gross features of the eye and ear and certain structures of some of the other receptors.

INTRODUCTION

The special sense organs are the eye and ear; tactile organs of pain, touch, and pressure; olfactory receptors of the nose; and taste buds on the tongue. The structure of the eye is readily apparent through dissection, that of the ear to a lesser degree. Organs of taste (gustatory) and smell (olfaction) and tactile sense organs are most readily observed through a study of slides, models, and drawings.

EYE
(Figures 10.1, 10.2)

Start the incision at the external corner of the eye where the two eyelids make contact. Cut around the entire eye and remove the upper and lower lids. Observe the thin mucous membrane, the *conjunctiva,* which covers the eyeball and is reflected onto the undersurface of the eyelids. In the medial corner of the eye, there is a white half-moon structure, the *nictitating membrane* or third eyelid. Remove the mucous connective tissue that lies between the eyeball and the surrounding bony orbital cavity, then dissect away the muscles and part of the bone surrounding the eye. Take care not to destroy the thin straps of muscle attached to the surface of the eyeball. Look on the ventral surface of the eyeball and observe two of the seven *ocular* muscles, the *inferior rectus* proceeding into the back of the orbital cavity, and the *inferior oblique* whose origin is on the cranial wall of the orbit. Push the eyeball forward and observe the *lateral rectus* muscle. The *superior rectus* and the *superior oblique* can be seen by moving the eyeball ventrally. Push the eyeball caudally and identify the *medial rectus.* Now cut the ocular muscles as far as possible from their insertion on the eyeball, clip through the *optic nerve* at the caudal end of the eyeball, and remove the eye from the orbital cavity. With the eye now free notice the conelike mass of muscle surrounding the optic nerve. This is the *ocular retractor* or *retractor bulbi* and serves to pull the eyeball back into the orbital cavity.

Using a razor blade or sharp scalpel, make a median sagittal incision through the entire eye and place the two halves in a dish of water. Using figure 10.2 identify the three coats that compose the eye. The external layer is white and opaque and is the *sclera,* the middle dark layer is the *choroid* coat, and the innermost layer is the *retina.* The large white structure within the eye is the *crystalline lens.* At the front of the eye is the transparent *cornea.* Surrounding the lens like a curtain is the colored portion of the eye called the *iris.* This is a cranial extension of the choroid. In the center of the iris is an opening, the *pupil.* Lying between the front of the lens and the cornea is a small space, the *anterior chamber,* and behind the iris is another smaller space, the *posterior chamber.* These spaces are filled with a fluid called *aqueous humor.* The large space caudal to the lens contains a clear gelatinous fluid called the *vitreous humor* or *vitreous body.* At the point of entry of the optic nerve into the caudal part of the eye is a region called the "blind spot" since it contains neither rods nor cones, which are the visual receptors of the eye.

EAR
(Figures 10.3, 10.4, 10.5)

The ear is composed of three parts, the *external, middle,* and *internal* ear. The external ear includes the *pinna* or *auricle* and the *external auditory meatus;* the *middle* ear consists of the *tympanic* or middle ear cavity in which are located three small bones, called the *ossicles.* These are the *malleus, incus,* and *stapes.* The external canal is separated from the tympanic cavity by the *tympanic membrane.* The three ossicles are attached to each other with the malleus in contact with the tympanic membrane, the incus next in line, and with the stapes fitted into the *fenestra vestibuli* or *oval window* in the wall of the *bony labyrinth.*

The inner ear has a canal-like system, the *bony labyrinth,* surrounding a *membranous labyrinth.* A fluid called *perilymph* fills the bony labyrinth; the membranous labyrinth contains fluid called *endolymph.* On its medial side, the vestibule is continued as the *cochlea,* a coiled snail-like structure containing the *Organ of Corti.* This structure possesses nerve receptors, the *hair cells,* which transmit nerve impulses to the auditory nerve. Arising from the vestibule opposite the cochlea are three *semicircular canals;* these are involved in the maintenance of equilibrium of the animal.

Cut off the pinna close to the head and identify the external auditory meatus which will be seen proceeding ventrally from the surface. Remove the skin and muscles surrounding the canal as well as the bone in the same area. Continue the dissection until the tympanic membrane is exposed at the end of the canal. The middle and inner ear structures will be more easily studied by the use of models and demonstrations.

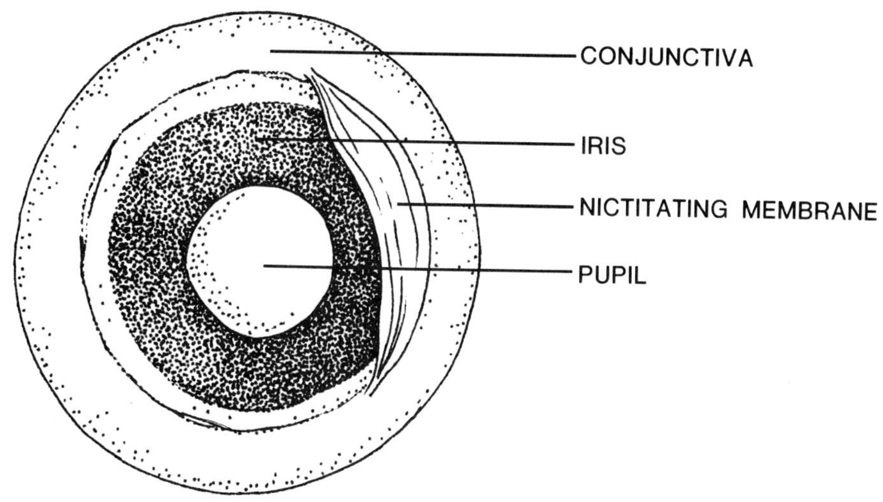

FIGURE 10.1.   Eye of fetal pig, frontal view.

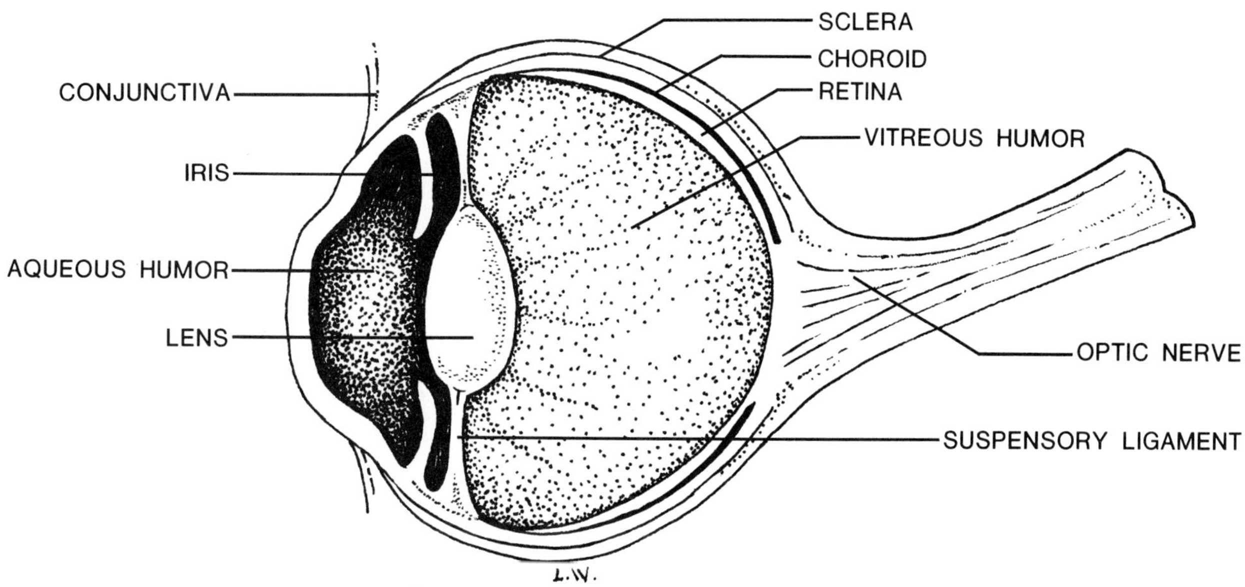

FIGURE 10.2.   Eye of fetal pig, sagittal section.

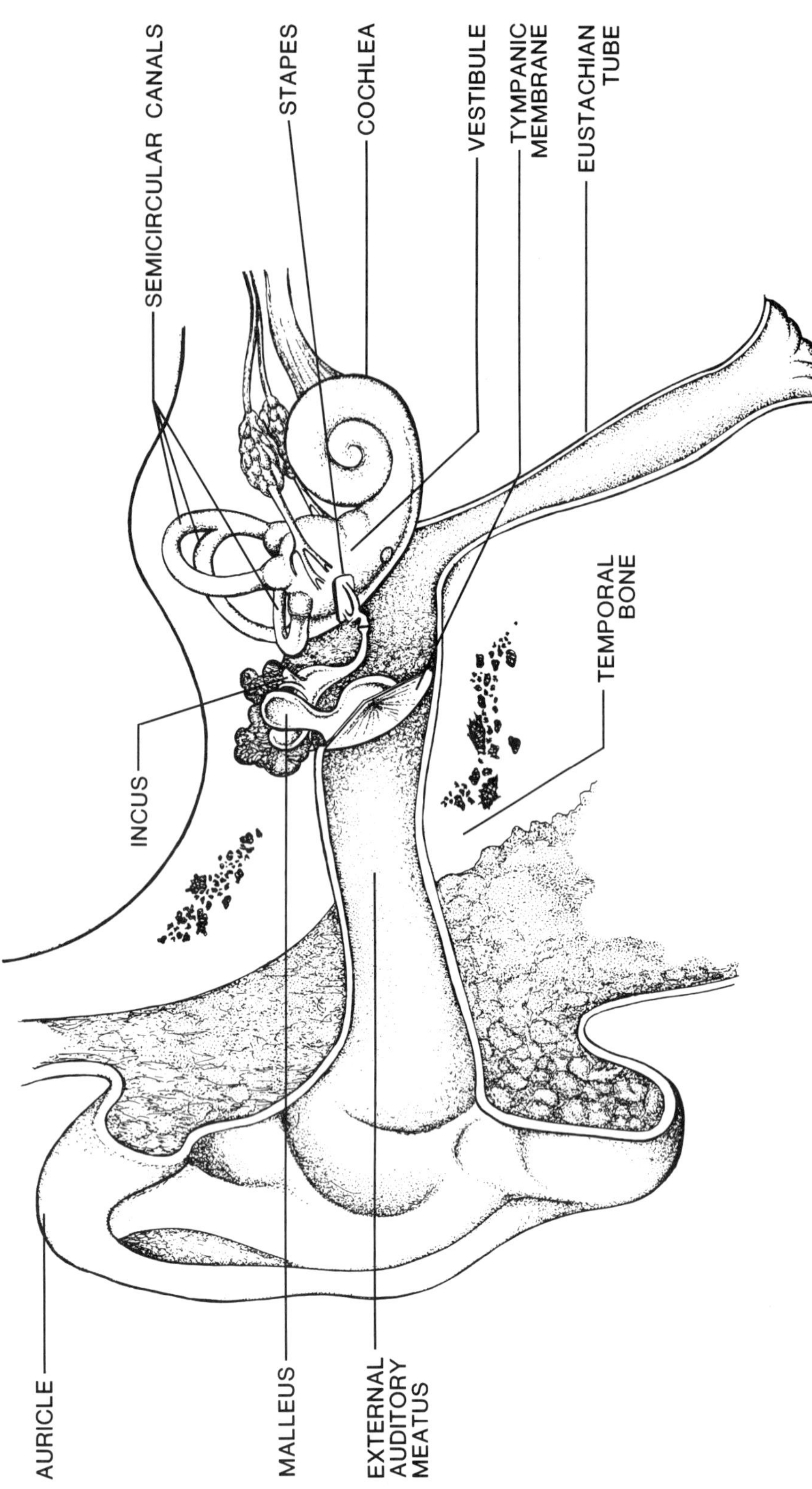

FIGURE 10.3. Structure of the inner ear. (By permission of John Raynor.)

FIGURE 10.4. Structure of the inner ear. (By permission of John Raynor.)

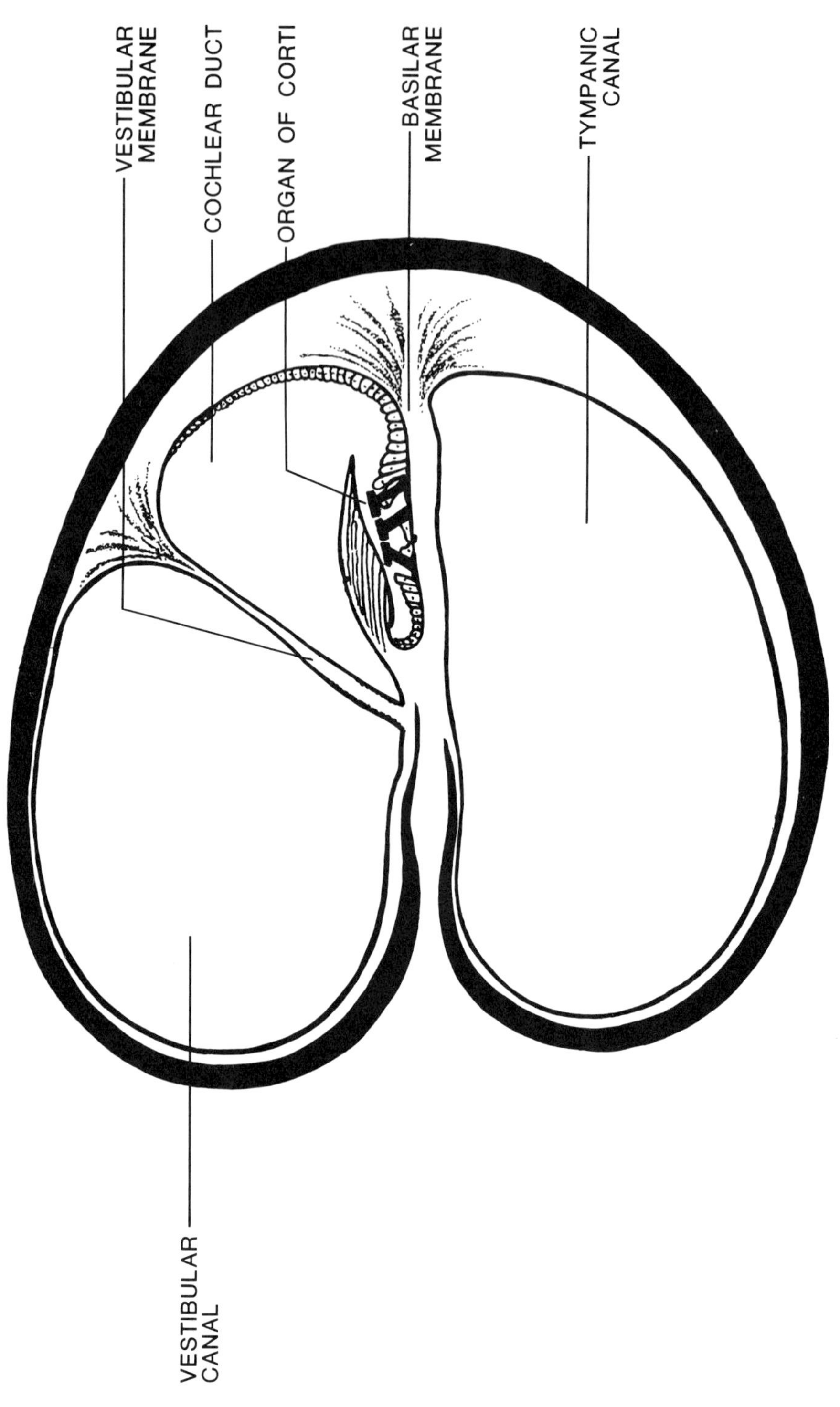

FIGURE 10.5. Structure of the cochlea. (By permission of John Raynor.)

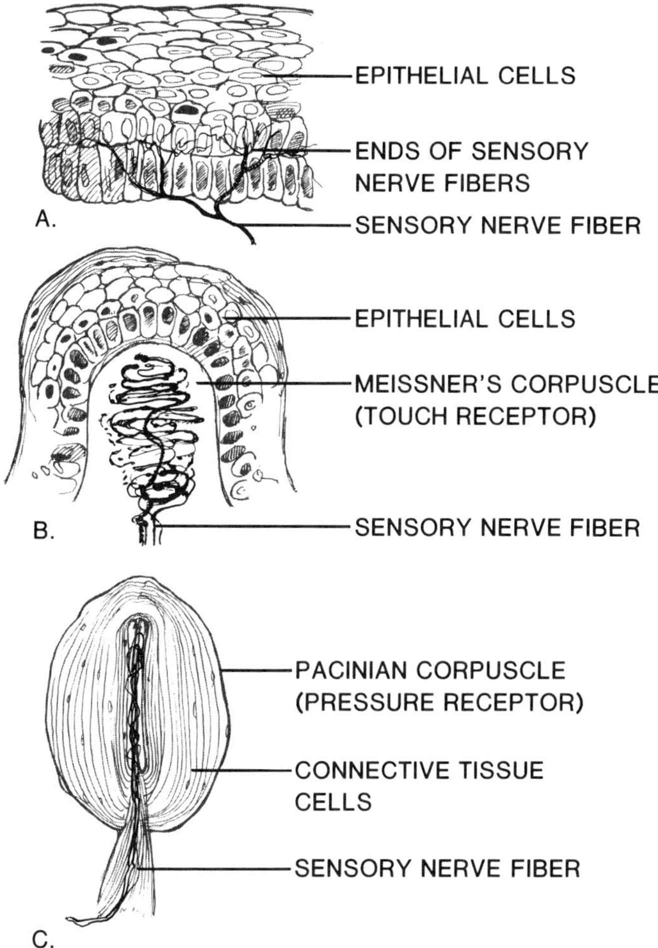

FIGURE 10.6. Touch and pressure receptors include (*a*) free ends of sensory nerve fibers; (*b*) Meissner's Corpuscles; (*c*) Pacinian Corpuscles.

TACTILE SENSE ORGANS
(Figure 10.6)

Figure 10.6 illustrates some of the organs of special sense concerned with (a) pain, (b) touch, and (c) pressure. These are known as *exteroceptors* since they react to stimuli received from the external environment.

Pain receptors are in the form of naked nerve fibers in the deep layers of the epidermis.

Meissner's Corpuscles function as end-organs sensitive to light touch. They are located in the dermis of the skin and are organized as a series of small vertical plates with wide distribution throughout the body such as in the skin of the arms, hands, and feet.

Pacinian Corpuscles are stimulated by deep touch or pressure. In appearance they are more compact than Meissner's Corpuscles, being composed of layers of closely applied connective tissue cells. These receptors are present in the subcutaneous layer of the skin of the palm of the hand and sole of the foot.

OLFACTION
(Figure 10.7)

Olfactory receptor cells are situated high in the nasal chamber on either side of the nasal septum. The basic component of this sense organ is composed of olfactory receptor cells whose free ends are continued as small hairs or *cilia* which react to incoming odors.

GUSTATORY SENSE
(Figure 10.8)

Located over certain areas of the surface of the tongue are small raised structures, the *papillae*. There are four types of papillae but only two, the *circumvallate* located on the posterior surface of the tongue and the *fungiform* at the apex, contain most of the taste buds. Taste buds are ovoid, spherical, or barrel-shaped in appearance each with a small pore opening on the surface of the tongue and with a taste hair projecting from the opening.

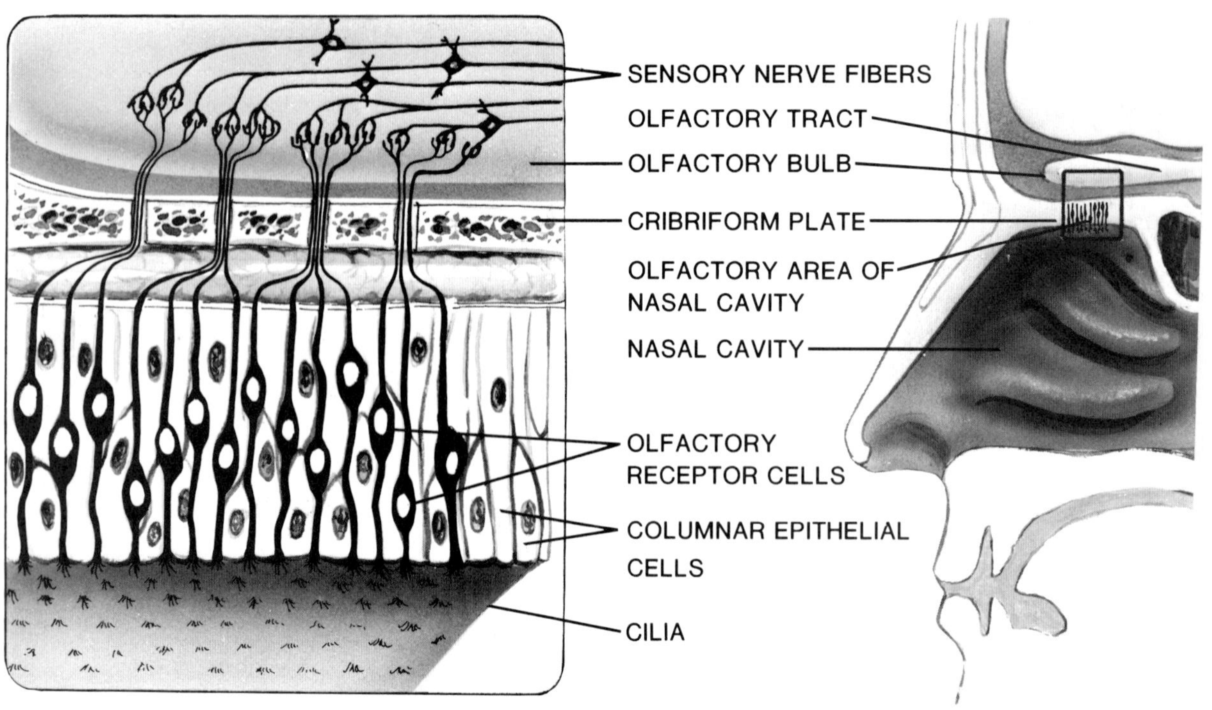

FIGURE 10.7. The olfactory receptor cells, which have cilia at their distal ends, are supported by columnar epithelial cells.

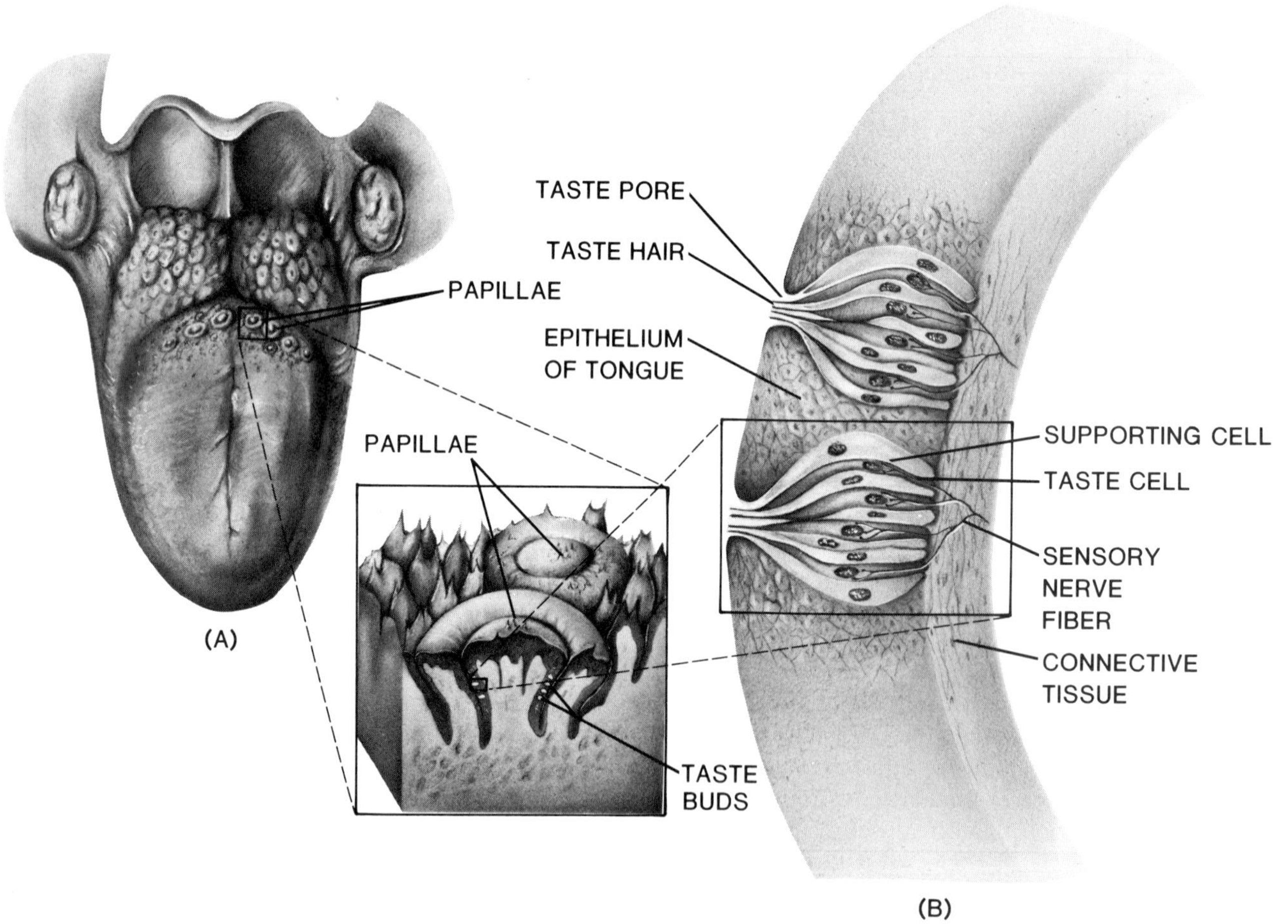

FIGURE 10.8. (*a*) Taste buds on the surface of the tongue are associated with nipplelike elevations called papillae; (*b*) a taste bud contains the taste cells.

Appendix 1
The Sheep Heart

(Figures H.1a, H.1b, H.2a, H.2b)

The sheep heart will have been removed from the thoracic cavity and the mediastinal membranes may or may not have been removed. If the pericardial membranes are still intact, review the arrangement of the thoracic coelomic membranes and then carefully separate the membranous walls of the mediastinal septum. The outermost membrane lines the thoracic (= pleural) cavity but is usually regarded as the outer layer of the *pericardia*. By tracing the evolution of these membranes, it is clear however that this outer membrane is actually the ventral mesentery of the thoracic cavity. Cut open the outer layer of the pericardium and note the fat and blood vessels and any glandular tissue between the inner and outer layers of the pericardium. Now remove the inner layer of the pericardium. This is the true parietal pericardium. The visceral pericardium adheres to the heart surface. The blood vessels and fat that you see on the surface of the heart are actually beneath the thin visceral pericardium and the cavity between the visceral and parietal pericardial membranes is filled with pericardial (= coelomic) fluid. The pericardial fluid reduces friction and cushions the violent movements of the heart.

VENTRAL SUPERFICIAL VIEW OF THE HEART

Atria (singular, *atrium* = a vestibule) are the two small sacs on either side of the cranial portion of the heart. The free edge of each atrium (opposite the entrance of the large veins) is termed the auricle (= little ear). The larger, caudal portion of the heart consists of two *ventricles* that superficially appear to be a single, large muscular structure. Ventrally the main branch of the large *left coronary artery* and the *great coronary vein* (beneath the artery) lie in the *interventricular paraconal groove* and thus mark the division between the right and left ventricles (fig. H.1a). Branches of the vessels continue to the right side in the *coronary groove* between the atria and ventricles and medial to the *pulmonary arch*. A small *marginal branch* occurs on the left margin of the left ventricle. Use the left coronary artery as a guide to the interventricular septum (which lies beneath it) when dissecting the heart.

DORSAL SUPERFICIAL VIEW OF THE HEART

The large vessels of the heart are best seen on the dorsal side (fig. H.2a). The cranial and caudal *vena cava* enter the right atrium and the *coronary sinus* enters the right atrium from the left, between the cranial and caudal vena cava. The *pulmonary veins* enter the left atrium, two from the right lung and two from the left lung. The two pulmonary veins from the left lung usually join before entering the atrium and thus may appear to be a single vessel. The cut ends of the *pulmonary* and *aortic arches* are cranial to the veins. The pulmonary arch will be seen in the dissected ventral view of the heart.

The *right coronary artery* and the *middle cardiac vein* occur together in the *interventricular subsinosal groove* and mark the division between the right and left ventricles on this aspect of the heart. Near the apex of the ventricles an *anastomosing* branch of the right coronary artery passes to the ventral surface of the heart to join the left coronary artery. The left and right coronary arteries originate in the right and left ventral cusps of the semilunar valve of the aorta.

The *coronary groove* is the division between the atria and the ventricles. The *great coronary vein* originates near the apex of the ventricles, passes through the interventricular paraconal groove to the coronary groove where it receives the *middle cardiac vein* and marginal veins from the ventricles and smaller veins from the atria. The great coronary vein enlarges to become the *coronary sinus* before opening to the right atrium.

DISSECTED VIEWS OF THE HEART

Ventral

With a pair of scissors, cut off the medial half of the right auricle and atrium to expose the interior as in figure H.1b. Internally, the atrium is a large chamber, but the auricle has muscular *trabeculae* (= pectinate muscles) extending between its inner and outer walls. Blood enters the right atrium from the cranial and caudal vena caval veins, which have drained blood from cranial and caudal parts of the body and from the coronary sinus, which receives blood from the wall of the heart itself. The left atrium drains blood from the lungs via the pulmonary veins.

In cross section the right ventricle is "half-moon shaped" and the left ventricle is round. This difference in appearance is due to the difference in thickness of the muscular walls. The right ventricle exerts enough force to drive the blood to the lungs (and back) but the left ventricle must pump the blood to all parts of the body. With a sharp scalpel cut through the wall of the right ventricle

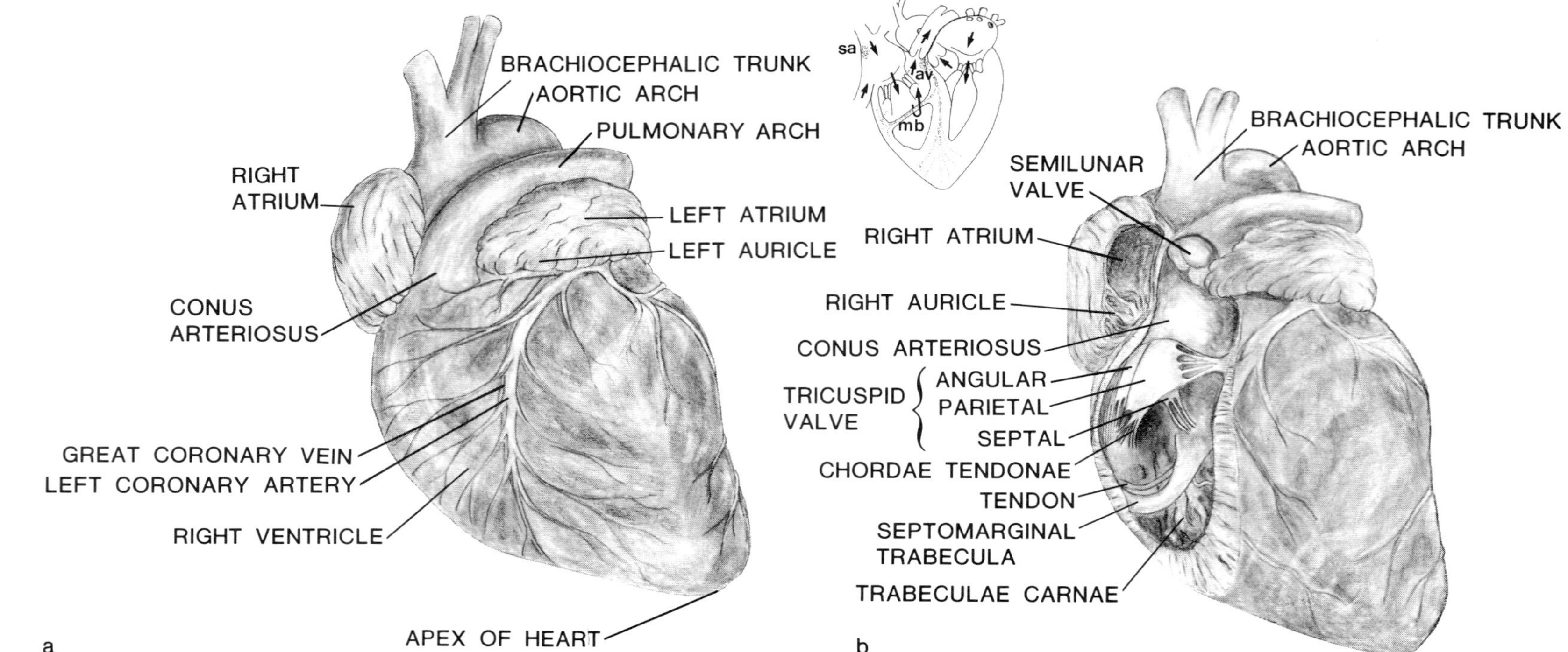

FIGURE H.1a.   Ventral superficial view of the sheep heart. The pericardial membranes, including the visceral pericardium and the fat that normally covers the superficial blood vessels, have been removed. In ventral views, the heart is shown as if the animal were laying on its back facing the reader. Therefore, the left side of the heart is to the reader's right and the right side of the heart is to the reader's left. b) Ventral dissected view of the sheep heart. The medial tip of the right atrium and auricle has been cut off with scissors, and the right ventricle has been carefully dissected to avoid cutting the papillary muscles and chorda tendinae supporting the tricuspid valve. The left ventricle has not been dissected on this side because a dorsal dissection of the left ventricle gives better access to the aortic semilunar valve and there is less danger of cutting the bicuspid valve in a dorsal dissection. The inset diagram is a ventral view of the dissected heart with arrows to indicate the flow of blood through the heart and the position of the heart conducting system (stippled areas). SA = sinoatrial node (pacemaker); MB = moderator band; AV = atrioventricular node.

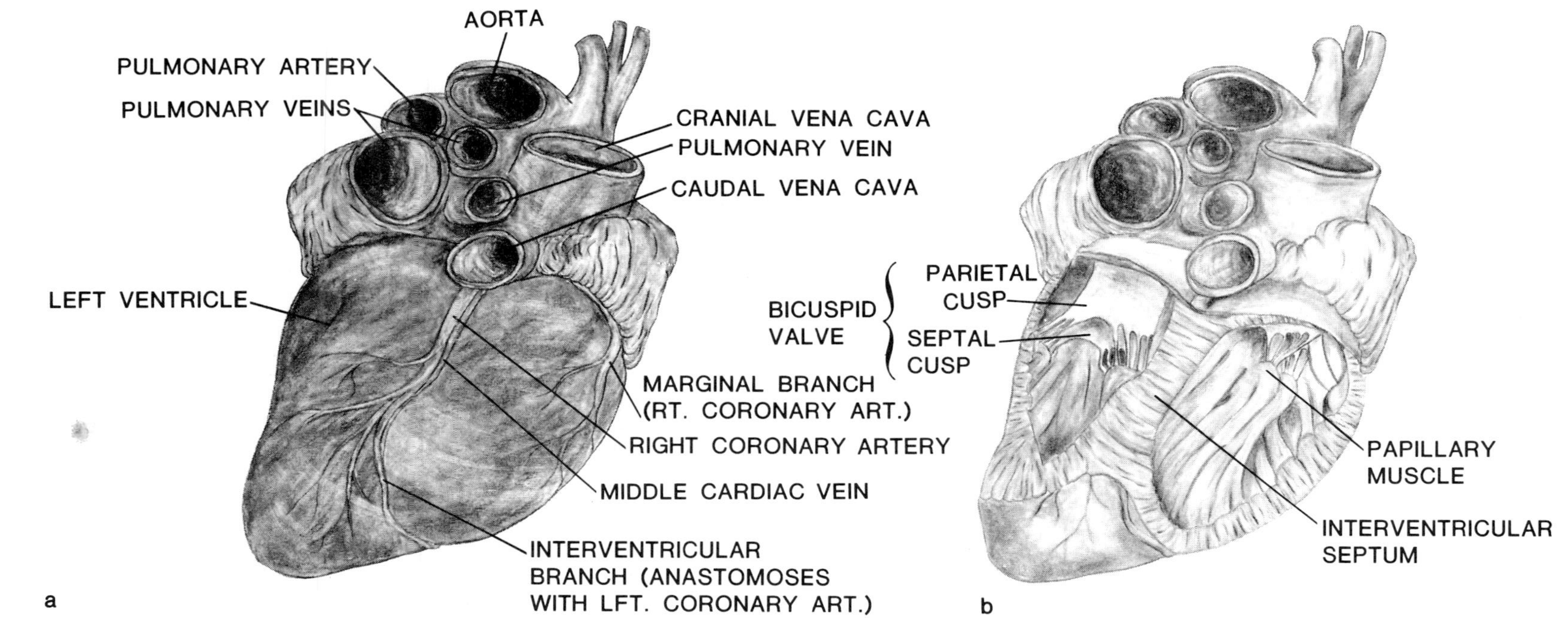

FIGURE H.2a. Dorsal superficial view of the sheep heart. The pulmonary veins are cut near their entrance to the left atrium and the vena cavae are cut close to their entrance to the right atrium. The right coronary artery lies in the interventricular subsinusoidal groove between the right and left ventricles. The marginal branch of the right coronary artery is on the right ventricle.
b) Dorsal dissected view of the sheep heart. The interior of the left ventricle is best seen from a dorsal view. The bicuspid (atrioventricular) valve surrounds the opening from the left atrium to the left ventricle. The lateral flap is nearest the outer wall of the ventricle and the septal flap is nearest the interventricular septum.

just to the right (your left) of the paraconal groove and the left coronary artery on the ventral surface of the heart. Start your cut at the apical end of the right ventricle but note that the right ventricle does not reach the apex, so confine your cut to the right side of the left coronary artery. Before cutting all the way to the coronary groove go back to the beginning of your incision and cut laterally along the right margin of the ventricle. This will provide a flap that you can lift up thus allowing you to observe and avoid the internal structures as you continue to cut and remove the ventral wall of the right ventricle. Your dissection should now appear similar to figure H.1b.

Note the muscular bands on the inner wall of the ventricles; these are *trabeculae carneae*. The larger muscle papillae with cords of tendon attached to them are called *papillary muscles* and the tendons are *chordae tendinae*. The cranial ends of the chordae tendinae are attached to tendinous flaps of the *tricuspid valve* on the right and to the *bicuspid* or *mitral valve* on the left side. Blood passes to the pulmonary artery from the right ventricle and into the aorta from the left. One or more trabeculae extend from the interventricular septum to the base of the papillary muscle supporting the lateral flap of the tricuspid valve. These trabeculae are called *septomarginal trabeculae* and contain branches of the interventricular conducting tract called *moderator bands* in addition to cardiac muscle fiber.

Both the pulmonary artery and the aorta are guarded by *semilunar valves*. Each valve consists of three cup-shaped flaps with the open end of the cup directed away from the heart. This arrangement allows the blood to flow away from the ventricles, but if blood were to flow toward the heart, the cups fill and close the opening between the ventricle and artery. Notice that the pulmonary semilunar valve is cranial to the level of the atrioventricular tricuspid valve and the wall of the ventricle in this area is similar to the wall of the pulmonary arch. This portion of the ventricle leading to the pulmonary semilunar valve is called the *conus arteriosus*.

Dorsal

Use the right coronary artery and the interventricular subsinuosal groove to orient the interventricular septum and begin your cut (with a sharp scalpel) on the left side just cranial to the apex of the heart. Again make your first cut just to the left of the coronary artery and toward the coronary groove. Make a second cut from the beginning of the first cut along the left lateral margin. Be careful to avoid cutting the papillary muscles supporting the chordae tendinae of the bicuspid valve. Remove the wall of the left ventricle by cutting transversely just below the coronary groove. You may also wish to remove the dorsal wall of the right ventricle as in figure H.2b. If so, be sure to leave a marginal band intact to support the angular flap of the tricuspid valve.

The lateral flap of the *bicuspid* valve is now exposed in the left ventricle. Beneath the lateral flap is the *septal* flap. Both flaps are held in place by chordae tendinae attached to papillary muscles. The *aortic semilunar* valve is just cranial to and between the lateral flap of the bicuspid and the dorsal wall of the ventricle. You should be able to locate the valve with a blunt probe. In order to see the valve you will need to dissect away most of the outer portion of the *aortic arch*. Before doing this, identify the *innominate* (= *brachiocephalic*), *subclavian,* and *carotid arteries.*

HEART BEAT

The sequence of heart contraction begins with the right atrium and is immediately followed by contraction of the left atrium. Contraction of the atrial muscles and constriction of the atrial chambers is termed *atrial systole*. This constriction forces blood through the atrioventricular valves (bicuspid and tricuspid) into the two ventricles. Dilation of the ventricles (*diastole*) accompanies atrial systole. During *ventricular diastole* blood flows into the ventricles and through the semilunar valves into the aorta and the pulmonary artery.

At the end of ventricular diastole, pressures in the atria, ventricles and great arteries (aorta and pulmonary artery) are equal. As the ventricles contract, *ventricular systole* begins and the pressure in the ventricles and great arteries increases while that of the atria decreases.

Muscles at the apex of the ventricles contract first and the contraction continues toward the atria. The atrioventricular valves prevent the flow of ventricular blood into the atria so the blood can only go into the great arteries through the semilunar valves.

At the end of ventricular systole, atrial diastole is also complete and atrial systole begins again. Semilunar valves prevent blood flow from the great arteries into the ventricles so the ventricles can only fill with blood from the atria.

Appendix 2
The Sheep Brain

(Figures B.1, B.2, B.3, B.4, B.5, B.6)

Sheep brains are available from biological supply companies as completely removed from the cranium but with the dura mater in place; removed from the braincase but sectioned into right and left halves; not removed from the braincase but with the cranial bones cracked to facilitate removal; cut in half sagittally and not removed from the cranium; or with the entire head preserved.

Only minimum instructions for dissection are presented here. If the student is required to remove the brain from the cranium the instructor should provide additional instructions.

After skinning, remove the temporalis and muscles of the occipital region. Use bone snips and cut away the bone of the lambdoidal ridge. This is the thickest bone area roofing the cranium. Be very careful not to cut into the brain. Continue cutting with the bone snips and chip off the remaining roof of the cranium. Leave the eyes intact for future study. The removal of the brain from the cranium is an extremely difficult and delicate operation. Cut the cranial nerves, leaving as much of the nerve attached to the brain as possible. Be very careful to loosen the pituitary gland from its socket (*sella turcica*) in the floor of the cranium.

GENERAL DESCRIPTION

The brain of mammals is divisible into anatomical regions corresponding to the major embryonic portions. The major categories, their subdivisions, and the major anatomical features for these divisions are as follows:

    I. PROSENCEPHALON
        Rhinencephalon Olfactory bulbs and tracts, Olfactory lobe (= *Amygdaloid body*)
        Telencephalon Basal nuclei, Cerebral hemispheres
        Diencephalon Hypothalamus, Thalamus, Epithalamus
   II. MESENCEPHALON
        Mesencephalon Colliculi and crus cerebri
  III. RHOMBENCEPHALON
        Metencephalon Cerebellum and Pons
        Myelencephalon Medulla oblongata

These terms may be used to refer to all of the structures in a particular region or to aid in locating a structure in the brain.

DORSAL VIEW OF THE BRAIN

Meninges are the membranes that cover the brain. Mammals have three layers of membranes; the *dura mater* is the outermost of the three. It is folded between the cerebral hemispheres as the *falx cerebri* and between the cerebrum and cerebellum as the *tentorium cerebelli*. The innermost layer is the thin *pia mater,* which adheres to the surface of the brain. Between the dura mater and pia mater is a network of thin delicate fibers called the *arachnoid* layer. The spaces between the arachnoid fibers are the *subdural* and *subarachnoid* spaces. Most of the blood vessels of the brain occur in the subarachnoid space, but the large venous sinuses adhere to the inner surface of the dura mater and arachnoid villi may project into these sinuses.

Carefully cut and remove the dura mater, if this has not been done previously, and note the pattern of blood vessels on the surface of the cerebral hemispheres.

Olfactory bulb (*bulbus olfactorius*) is at the extreme rostral end of the brain. The olfactory nerves terminate here after passing through the fenestrated ethmoid plate from the nasal epithelium. Some of the olfactory nerves remain attached to the bulb giving it a brushlike appearance. Notice that the olfactory bulb is a two-lobed structure in the sheep and there is a dorsal and ventral grouping of the *olfactory nerves.* The *olfactory gyrus* consists of fiber (nerve) tracts between the olfactory bulb and the *hippocampus.* These centers are concerned with the sense of smell.

The sense of smell is a very primitive sense. Nearly all the telencephalon of early vertebrates (i.e., dogfish shark) is devoted to this sense, and the cerebral hemispheres make up a smaller part of their telencephalon.

Cerebrum consists of two halves or hemispheres divided by a deep longitudinal groove, the *longitudinal cerebral fissure.* The cerebrum of most mammals has a folded surface with fissures or *sulci* indenting the surface and ridges or *gyri* projecting outward. The surface layer of the cerebrum is a cortical layer of gray matter that is increased by the surface folding. Since gray matter is the area where nerve cell bodies and synapses occur, the greater the amount of gray matter the greater the number of synapses and the greater the number of pathways an impulse may travel. Because of this relationship it is often believed that the amount of "gray matter" in the cerebrum is related to "intelligence" of mammals. Although this concept is debatable it does point out the importance

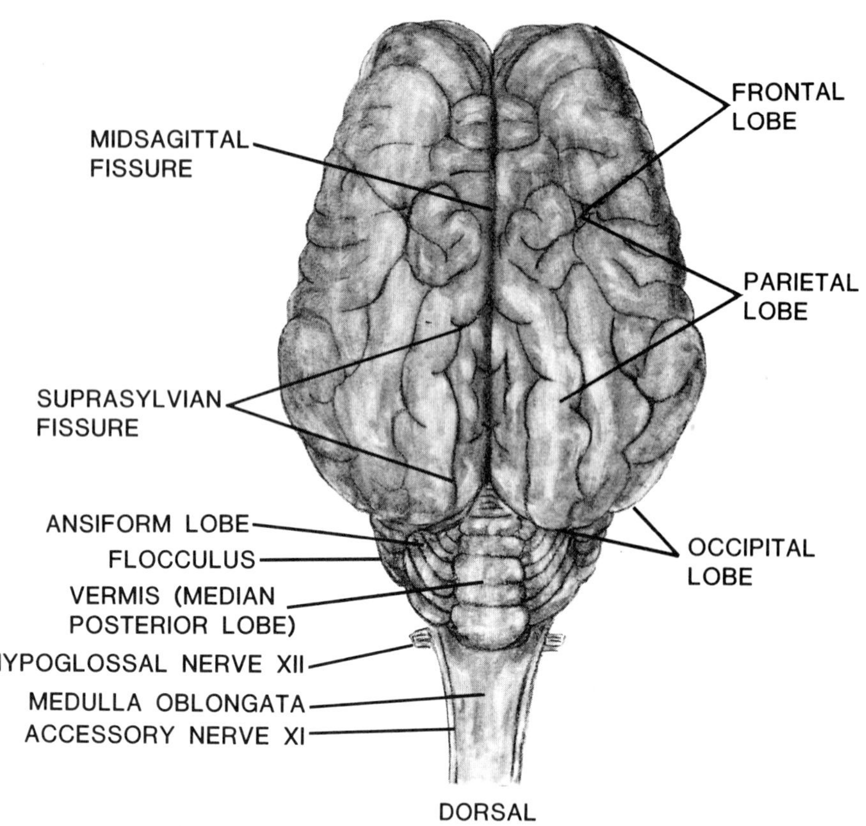

FIGURE B.1. Dorsal view of the sheep brain with some of the lobes of the cerebrum and cerebellum labeled. Compare with figure B.3.

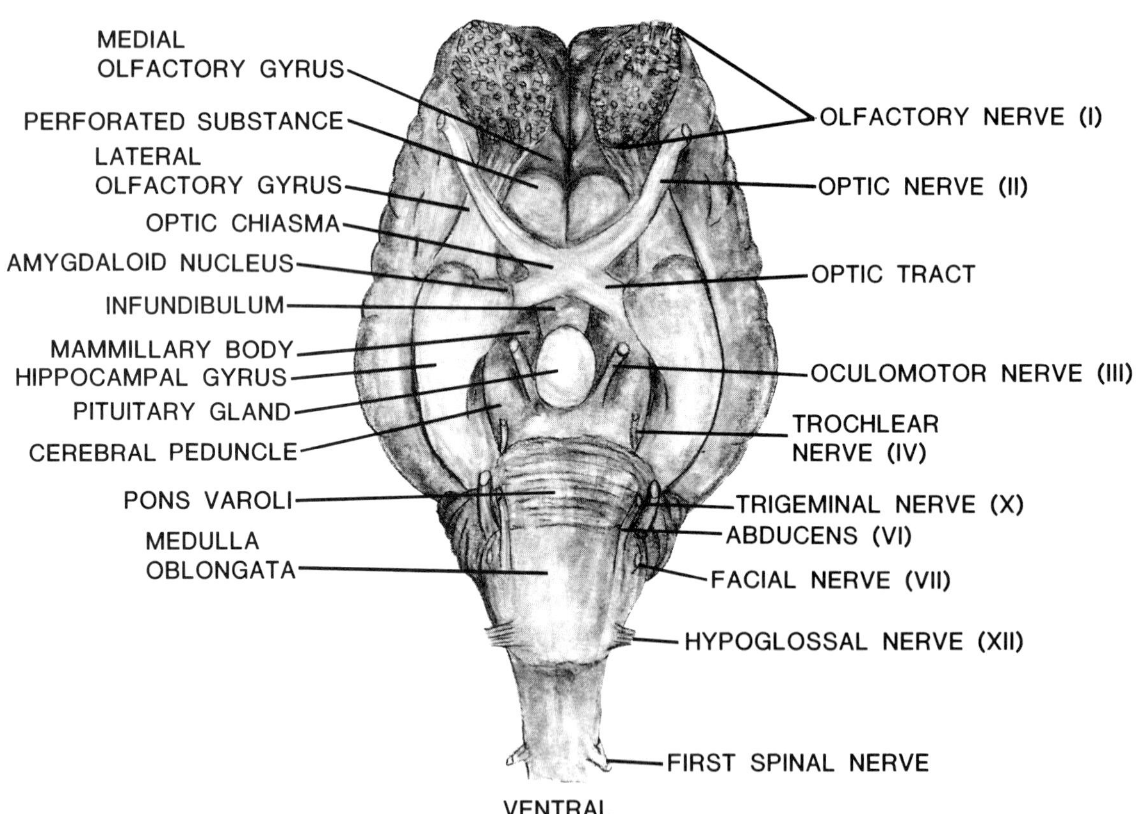

FIGURE B.2. Ventral view of the sheep brain showing the position of the cranial nerves.

of the cerebral folding and of the assignment of names to the various gyri and sulci (or fissures). The surface of the cerebrum in this dorsal view is all gray matter. The difference between gray and white matter will be seen in the sagittal or cross-sectional views. The deeper fissures help to divide the hemispheres into *lobes.* The major fissures and lobes of the sheep cerebrum are best seen in dorsal and lateral views.

The *corpus callosum* is a band of fibers on the floor of the longitudinal fissure. This band of nerve fibers connects the gray matter areas of the cerebral hemispheres. In order to see these fiber tracts, gently spread the cerebral hemispheres apart with your thumbs while holding the whole brain in your hands. A cross section of this structure will be seen in sagittal view.

The *pineal body* (corpus pineale), is a stalked gland arising from the midbrain just rostral to the corpora quadrigemina. This will also require that you spread the cerebral hemispheres apart in order to see the structure. You may also need to press the cerebellum back, very carefully, in order to see the pineal. The pineal body releases the hormone melatonin in darkness but the function of this hormone in mammals is unknown. It is suspected of having an inhibiting effect on the gonads.

Corpora quadrigemina are composed of two pairs of lobes between the cerebrum and the cerebellum. The rostral pair are the *rostral colliculi* and serve as optic reflex centers. The caudal pair are the *caudal colliculi* and serve as otic (auditory) reflex centers.

Cerebellum is the highly convoluted portion of the brain caudal to the corpora quadrigemina. It consists of two lateral hemispheres, each with laterally projecting *folliculi,* and a median *vermis.* As in the cerebrum the deep fissures are sulci and the folds between sulci are gyri. The lateralmost portion of each hemisphere is extended as a flocculus. Ventral to the flocculi, the cerebellum is attached to the medulla and pons by three pairs of peduncles: *caudalis, medius,* and *rostralis.*

The cerebellum is principally a motor coordinating center. The hemispheres are primarily responsible for coordinating movements of the limbs and digits while the vermis has areas for coordinating movements of the spinal column, shoulders, and hips.

Medulla oblongata is the most caudal part of the brain and constricts imperceptibly into the spinal cord. A groove, the *sulcus medianus,* divides the medulla into right and left halves. This structure will be better seen in lateral and ventral views. The functional centers in the medulla include respiration, heart rate, and blood pressure but it also plays an important role in regulating or mediating sensory impulses, endocrine secretions, and general awareness.

VENTRAL ASPECT OF THE BRAIN

Most of the structures seen in this view were also seen in the lateral view of the brain, but the cranial nerves are more easily identified in a ventral view. Review the olfactory gyrus, hippocampus, and piriform area seen in the lateral view. In ventral view the medial bulge from the hippocampal gyrus is the *amygdaloid.*

Optic chiasma, crossing of the optic nerves between the caudal ends of the olfactory tracts. (See Chart of the Cranial Nerves for the *optic nerves.*)

Optic tracts are continuations of the optic nerve fibers from the chiasma over the lateral surface of the diencephalon to the *lateral geniculate body,* the occipital lobe of the cerebrum, and to the rostral colliculi of the corpora quadrigemina.

Tuber cinereum, a prominence caudal to the optic chiasma and just rostral to the mammillary bodies. This is the area of attachment of the hypophyseal infundibulum to the brain. If the pituitary gland was pulled away during preparation the tuber cinereum will be seen as the area around a small hole (opening to the third ventricle; see sagittal view).

Hypophysis or pituitary gland is an important gland situated within the sella turcica and surrounded by a meningeal fold which may prevent removal of the gland with the brain. The hypophysis has two distinct parts (1) the adenohypophysis derived from the embryonic pharynx and divisible into *pars intermedia* and a *pars distalis,* and (2) the *neurohypophysis,* an extension of the floor of the hypothalamus. The proximal portion of the neurohypophysis is the *infundibulum* or stalk, and the distal portion is referred to as the *pars nervosa* in the physiological literature. A small cavity, the cavum hypophysis may separate the pars intermedia and pars distalis of the adenohypophysis. The neurohypophysis has no secretory cells, but the axons of cells in the hypothalamus release their secretions from the neurohypophysis. These hormones are vasopressin or antidiuretic hormone, which increases reabsorption of water by the kidney and oxytocin, which stimulates milk ejection from the female mammary gland. The pars intermedia secretes melanophore stimulating hormone (MSH), which causes melanin formation by melanophores (pigment cells) in the skin. Although most mammals have a pars intermedia it is usually not present in adult humans. The pars distalis secretes six hormones: (1) thyroid-stimulating hormone (TSH), which stimulates thyroid gland cells, (2) adrenocorticotropic hormone (ACTH), which stimulates adrenal cortex cells, (3) follicle-stimulating hormone (FSH), which stimulates follicle growth in the ovary of females and spermatogenesis in the testes of the male, (4) luteinizing hormone

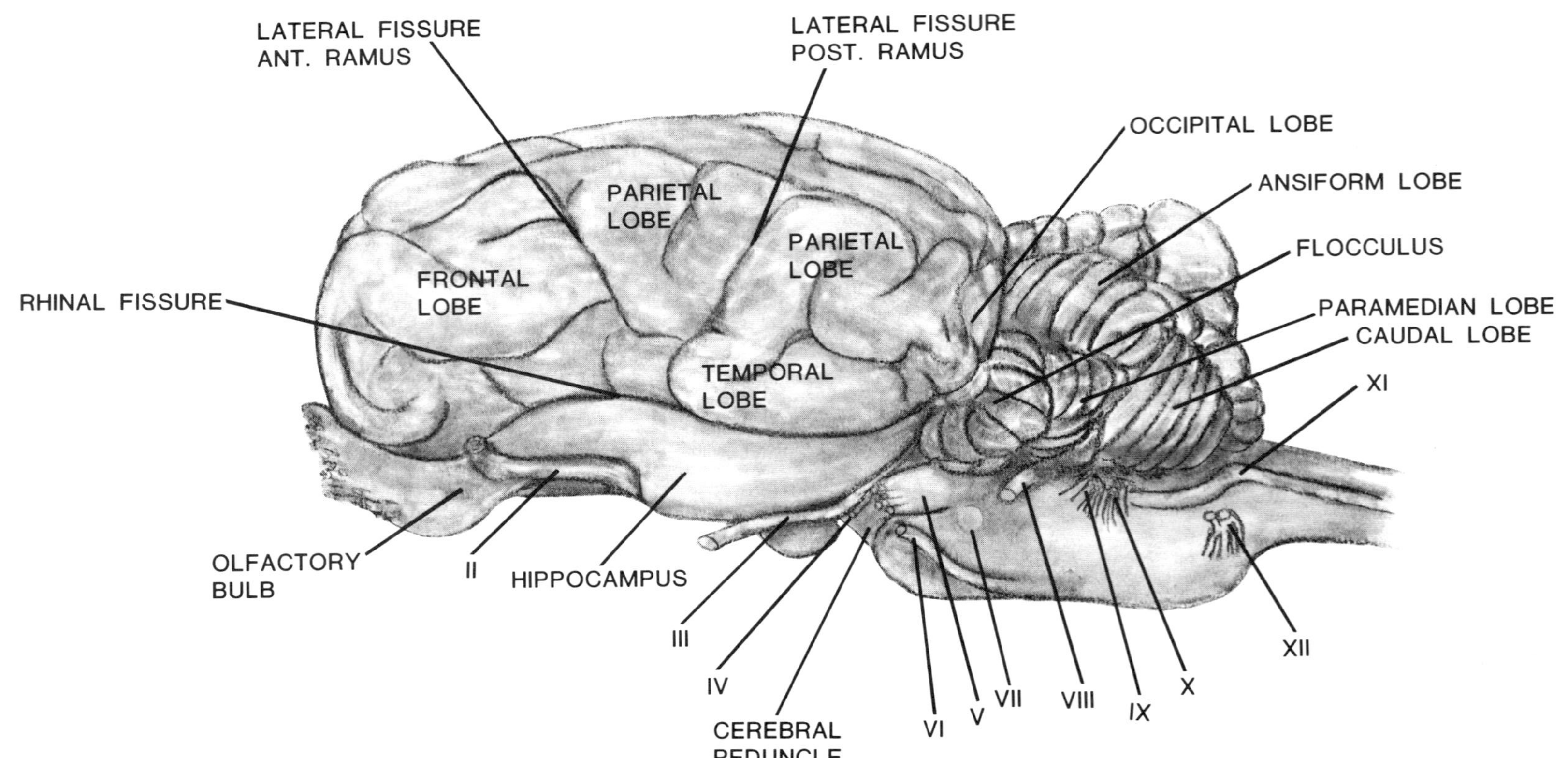

FIGURE B.3. Lateral view of the sheep brain illustrating the lobes and fissures of the cerebrum and cerebellum, and the lateral olfactory system and cranial nerves. Also see figure B.5.

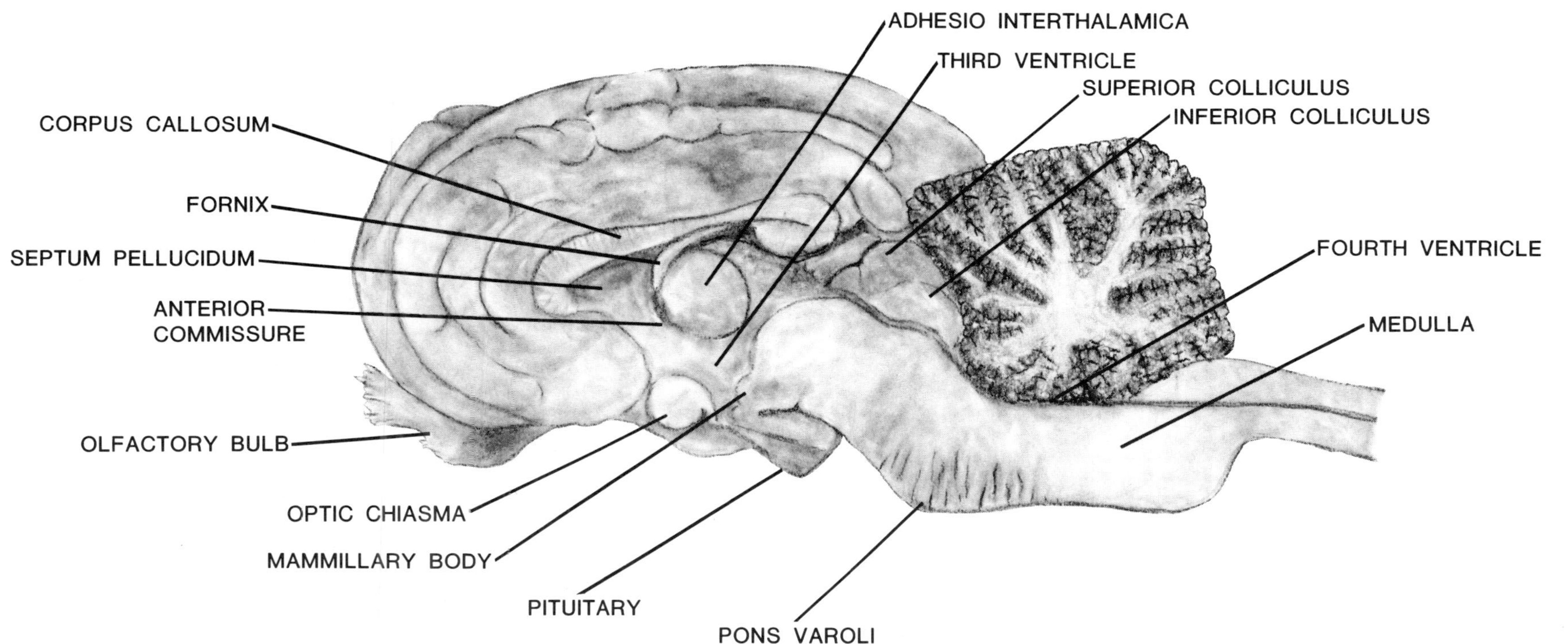

FIGURE B.4. Sagittal section of the sheep brain.

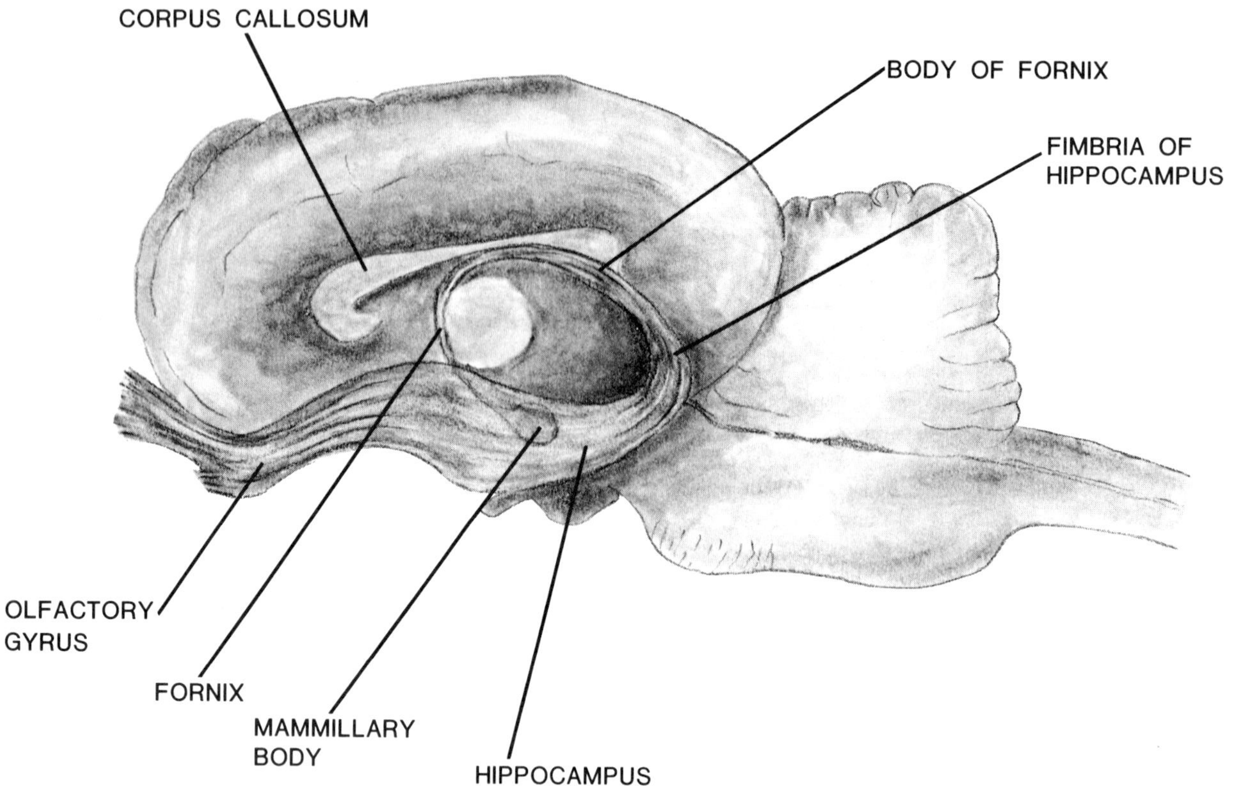

FIGURE B.5. Diagrammatic illustration of the major olfactory tract system of the sheep brain. The structures illustrated here extend from the lateral plane seen in figure B.3 to the midsagittal plane of figure B.4 and include areas between the two.

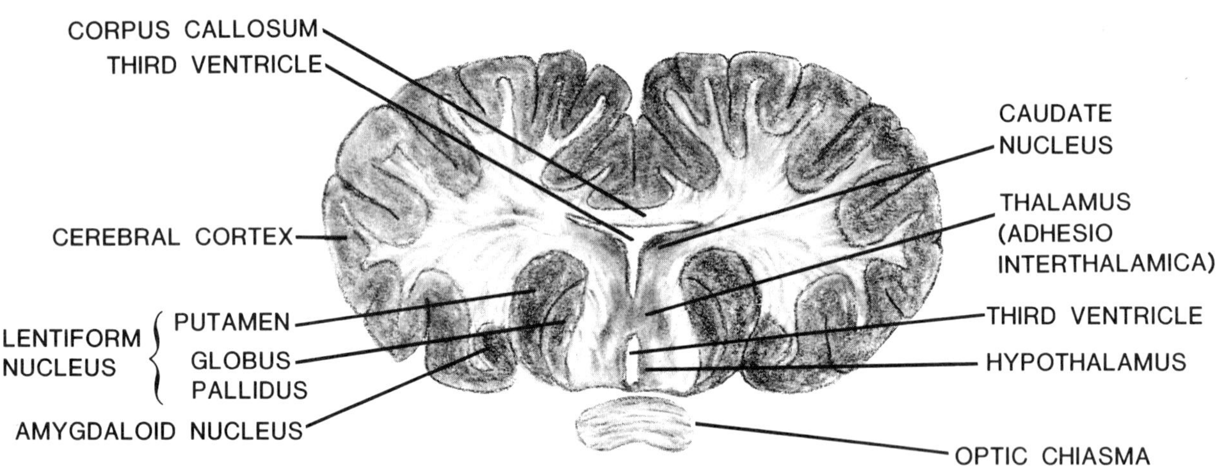

FIGURE B.6. Cross section of the sheep brain through the midbrain.

NAME	SUPERFICIAL ORIGIN ON THE BRAIN	ACTION	DISTRIBUTION
I Olfactory (multiple branches)	Olfactory bulb (with connections to olfactory, gyrus, hippocampus, etc.)	Sensory	Nasal epithelium
II Optic	Superior colliculi and thalamus (with connections to occipital lobe)	Sensory	Retina of the eye
III Oculomotor	Pedunculi cerebri	Motor	Dorsal, ventral, and medial recti and ventral oblique eye muscles
IV Trochlear	Caudal to inferior colliculi orbital fissure	Motor	Dorsal oblique eye muscle
V Trigeminal	Pons	Sensory and motor	Skin (ophthalmic) vibrissae (superior or maxillary) jaw muscles, tongue, teeth (inferior maxillary)
VI Abducens	Rostral medulla oblongata	Motor	Lateral rectus and retractor bulbi eye muscles
VII Facial	Medulla oblongata near V	Sensory and motor	Masticatory muscles
VIII Auditory	Medulla oblongata, caudal to VII (with connections to caudal colliculus and thalamus)	Sensory	Hair cells of inner ear
IX Glossopharyngeal	Medulla oblongata near X	Sensory and motor	Pharynx and tongue
X Vagus	Medulla oblongata caudal to VIII	Sensory and motor	Larynx, heart, lungs, diaphragm, and stomach
XI Spinal Accessory	Medulla oblongata and rostral end of spinal cord	Motor	Muscles of neck and pharyngeal viscera with vagus
XII Hypoglossal	Medulla oblongata caudal to X	Motor	Tongue muscles

(LH), stimulates ovulation and the formation of the corpus luteum in the ovary and the secretion of testosterone by cells in the testes, (5) prolactin, which stimulates milk secretion and maintains the corpus luteum in the rat ovary, and (6) growth hormone (somatotrophin), which stimulates growth.

Crus cerebri, fiber tracts flanking the infundibulum and connecting the cerebrum and medulla oblongata.

Pons, fiber tracts on the ventral, rostral end of the medulla connecting the hemispheres of the cerebellum.

Trapezoid body (corpus trapezoideum), a thickened body caudal and parallel to the pons on the ventral surface of the medulla.

Cranial nerves. There are twelve pairs of cranial nerves, which are described in the above chart. Learn these nerves by name, number, origin, and distribution.

LATERAL VIEW OF THE BRAIN

Olfactory tracts are broad bands of nerve fibers extending caudally and ventrally from the olfactory bulb to merge into the *hippocampal area* at the level of the *optic chiasma.* In the lateral view (fig. B.3), the olfactory band of fibers appears arched or bent and they are therefore referred to as the *olfactory gyrus* (from the Greek *gyros* = a turn).

Hippocampus is an olfactory association center with connections to other olfactory and to visual association centers. The major connections of the hippocampus are through the fibers of the fornix (see sagittal view) that extend from here to the *mammillary body.* The hippocampus is broader and more rounded than the olfactory gyrus. The laterally rounded portion of the hippocampus is termed the hippocampal gyrus and the olfactory and hippocampal gyri together may be termed the *pyriform area.* The fornix and mammillary body will be seen in the sagittal section of the brain.

Cerebral Hemispheres

Cerebral hemispheres were partly described in the dorsal view. Each hemisphere is divided into four major lobes by the fissures and branches of fissures (= ramus). The ventral, *rhinal fissure* separates the hemisphere from the olfactory and hippocampal gyri. A more or less vertical *lateral* fissure divides the hemisphere into rostral and caudal halves. This is also referred to as the *fissure of Sylvius,* and in the sheep there is a prominent caudal ramus that partially subdivides the *parietal* lobe and sets off the *temporal* lobe as a distinct portion. The rostral portion of the lateral fissure separates the *frontal* and parietal lobes. The *insula* is a discreet portion of the cerebrum between the frontal, parietal, and temporal lobes and above the rhinal fissure. The *occipital* lobe is not separated by distinct fissures but it includes the most caudal median gyrus.

The lobes of the cerebrum area usually associated with the broad functional categories that are listed here, but there is no attempt to list all the functions of the lobe nor to describe the interrelationships of the lobes. Furthermore, the experiments that have identified the functions listed below were performed mainly on cats and monkeys but not on sheep.

Frontal Lobe. The half of the lobe rostral from the lateral fissure is concerned with primary motor function in most mammals studied. Localized functional patterns correspond closely with localized sensory patterns of the sensory association cortex in the parietal lobe. Most motor functions are initiated here. The cortex rostral to the primary motor cortex has connections with the thalamus, and stimulation of this cortex in the monkey affects blood pressure, respiratory rate, and intestinal tract motility and (perhaps through the thalamus) emotional behavior.

Parietal Lobe. A narrow region caudally from the lateral fissure is the sensory association cortex receiving input from all parts of the body. Caudal to the sensory association cortex is a somatic (touch, heat, cold, pressure) area, and caudal to this is a cortex concerned with visual coordination and association.

Occipital Lobe. In most animals this area is associated with visual projection and association. Stimulation of this area in the monkey causes specific eye movements.

Temporal Lobe. In humans this lobe is concerned with speech, which includes production and association of speech with writing and vision. Although writing and speech are not performed by the sheep, voice and visual association with voice may be centered in this lobe.

Insula. Insula is associated with olfaction through the pyriform area and may also be involved in taste. This lobe is also thought to have a role in visceral motor function.

Cerebellum

Cerebellum is also convoluted with sulci and gyri similar to those of the cerebrum. It consists of two lateral hemispheres, which may be divided into *simplex, ansiform, follicular,* and *paramedian* lobes and a median *vermis,* which is also lobulated. Ventral to the flocculi, the cerebellum is attached to the medulla and pons by rostral, middle, and caudal peduncles. The hemispheres are responsible for coordinating movements of the limbs and digits while the vermis has areas for coordinating movements of the spinal column, shoulders, and hips. The hemispheres of the human brain are so large that they cover the vermis completely, which correlates with the greater digital coordination by humans.

SAGITTAL SECTION

Cut the brain in half longitudinally with a sharp scalpel through the sagittal fissure. You will need to spread the halves of the brain apart as you cut and you will need to make more than one slice since your blade is not long enough to cut through the brain in one stroke. Keep your cuts as close to the midline as possible. Locate the following:

> *Corpus callosum,* band of fibers connecting the cerebral hemispheres. The caudal portion is an oval mass, the splenium, and the rostral is deflected ventrally at the rostrum.
>
> *Fornix,* a fiber tract connecting the hippocampus (see figs. B.2 and B.3) with the mammillary body (corpus mammillare). The tract runs ventral to the third ventricle then turns caudally to the mammillary body. The connections of this tract with the hippocampus and indirectly with the olfactory tract are illustrated in fig. B.5, but there are several connections with other parts of the brain that are not illustrated.
>
> *Septum pellucidum* consists of two narrow membranes of neural tissue extending between the corpus callosum and fornix and separated by a narrow cavity. The septum pellucidum is located in the midline of the cerebrum and separates the lateral ventricles (= I and II) in the cerebral hemispheres.
>
> *Third* (III) *ventricle,* a cavity just caudal to the fornix which extends ventrally into the infundibulum and caudally to the level of the corpora colliculi.
>
> *Adhesio interthalamica,* in the center of the third ventricle and connecting the walls of the third ventricle. This has often been referred to as the massa intermedia. There are no fiber tracts in this body. The nuclei of the thalamus in this region bulge toward the midline on each side so the walls of the ventricle contact each other in the midline.
>
> *Rostral commissure,* a rostral fiber tract connecting

the basal nuclear areas of the right and left hemispheres near the rostral end of the fornix.

Lamina terminalis is the rostral boundary of the third
ventricle, ventral to the fornix.

Mesencephalic aqueduct, duct connecting the third
and fourth ventricles. This is also called the *aqueduct of Sylvius.*

Interventricular foramen, the connection between the
first, second, and third ventricles just caudal to the
fornix. Also called the *foramen of Munro.*

Fourth (IV) *ventricle,* the cavity of the medulla oblongata just ventral to the cerebellum.

Tegmen ventriculi, thin sheets of nervous tissue covering the fourth ventricle and contributing to the
tela choroidea.

Tela choroidea, sheets of neural tissue and blood vessels covering the roof of each ventricle (including
the tegman ventriculi of the fourth ventricle).

Flocculus, lateral extensions on either side of the cerebellum.

BRAIN CROSS SECTION

White matter represents fiber tracts or nerve processes, and gray matter consists of cell bodies with nuclei
and nerve synapses. Gray matter occurs in layers (cortex)
or in units termed nuclei. White matter usually occurs in
tracts. The following areas of gray and white matter should
be located on a section of the brain through the optic
chiasma and the adhesio interthalamica (massa intermedia). An additional, caudal section of the brain might
be made just caudal to the pineal and through the rostral
colliculus and the rostral end of the medulla.

Cerebral cortex is the gray matter forming a thin layer
on the outer surface of the cerebrum. Notice that the gray
matter follows the folds (sulci and gyri) of the cerebral
cortex. The cortex and the underlying white-matter fiber
tracts together form the pallium. The cells of the cortex
are arranged in strata of three to six layers. All tetrapod
vertebrates have a stratified cortex with at least three
layers. This three-layer cortex is thought to be the most
primitive, hence is named archicortex (archipallium with
its underlying white matter). The highest development of
six cortical layers occurs only in mammals and is called
the neocortex (neopallium).

The archicortex is the hippocampus (see figs. B.3 and
B.5). The archicortex and the paleocortex are part of the
rhinencephalon or "olfactory" brain.

The neopallium and the basal ganglia (corpus
striatum) form the nonolfactory portion of the telencephalon. The neocortex is located roughly on the dorsolateral
surface of the cerebral hemispheres. In the sheep brain
the surface of the cerebrum is folded into gyri and sulci,
thus increasing the amount of neocortex.

Corpus callosum is the large bundle of white fibers
connecting the gray matter of the two cerebral hemispheres.

Corpus striatum consists of the basal nuclei (*pallidum, putamen,* and *caudate*) and the fiber tracts of the
internal capsule. Phylogenetically the putamen and nucleus caudatus are the newest portions of the corpus
striatum and may be referred to as the neostriatum. The
pallidum is an older feature; the paleostriatum and two
additional nuclei, the *claustrum* and amygdaloid body
(corpus amygdaloideum), form the archistriatum or most
primitive of the subcortical gray matter. The archistriatum is separated from the corpus striatum by fiber tracts
of the external capsule.

Thalamus consists of several nuclei medial and mainly
caudal to the basal nuclei. Thalamic nuclei are the dorsal
half of the diencephalon or "between" brain.

Hypothalamic nuclei form the ventral part of the
diencephalon. Some of these nuclei are known to secrete
hormones. (See sagittal view.)

Internal capsules are fiber tracts connecting various
areas of the cerebral cortex with the lateral or medial geniculate body or with nuclei in the pons.

Epithalamus is a very small area dorsal to the thalamus and consisting of the *habenulae* (nuclei), *stria med-
ullaris* (fiber tracts), and *pineal body.*

SUGGESTED READING

Ranson, S. W., and S. L. Clark. 1959. The Anatomy of the Nervous
System: Its Development and Function. Tenth Edition. W. B. Saunders Company. Philadelphia. [Although this book is old and now out
of print it contains an excellent description of the sheep brain and
may be found in most large libraries.]

Credits

Figures 2.3, 2.4, 2.5, 2.6, 2.7, 2.8, 2.9, and 2.10 from Chiasson, Robert B. and Ernest S. Booth, *Laboratory Anatomy of the Cat,* 7th ed. © 1967, 1972, 1977, 1982, 1989, Wm. C. Brown Publishers, Dubuque, Iowa. All Rights Reserved. Reprinted by permission.

Figures 10.6, 10.7, and 10.8 from Hole, John W., Jr., *Human Anatomy and Physiology,* 3d ed. © 1978, 1981, 1984 Wm. C. Brown Publishers, Dubuque, Iowa. All Rights Reserved. Reprinted by permission.

Appendix 1 and Appendix 2 from Chiasson, Robert B., *Laboratory Anatomy of the White Rat,* 5th ed. © 1988 Wm. C. Brown Publishers, Dubuque, Iowa. All Rights Reserved. Reprinted by permission.

Index